D0429929

The St. Lucie Press/APICS Series on Resource Management

Titles in the Series

Applying Manufacturing Execution Systems
by Michael McClellan

Back to Basics:
Your Guide to Manufacturing Excellence
By Steven A. Melnyk
R.T. "Chris" Christensen

Enterprise Resources Planning and Beyond:
Integrating Your Entire Organization
by Gary A. Langenwalter

ERP: Tools, Techniques, and Applications
for Integrating the Supply Chain
by Carol A. Ptak with Eli Schragenheim

Integral Logistics Management:
Planning and Control of Comprehensive Business Processes
by Paul Schönsleben

Inventory Classification Innovation:
Paving the Way for Electronic Commerce and
Vendor Managed Inventory
by Russell G. Broeckelmann

Lean Manufacturing:
Tools, Techniques, and How To Use Them
by William M. Feld

Macrologistics Management:
A Catalyst for Organizational Change
by Martin Stein and Frank Voehl

Restructuring the Manufacturing Process:
Applying the Matrix Method
by Gideon Halevi

Supply Chain Management:
The Basics and Beyond
by William C. Copacino

Lean Manufacturing

Tools, Techniques, and How To Use Them

Dedication

This book is dedicated to my four sons —
Benjamin, Nathan, Jacob, and Samuel.

Thank you for never letting me forget that I am your dad.

Contents

Part IV. Case Studies

Preface

This book was written in order to give the general manufacturing practitioner a reference guide by which to lead the successful design and deployment of a lean manufacturing program. It is for those individuals who have either tried a lean manufacturing implementation and received undesirable results or have been working at it a while and do not really know what to do next. Over the years, I have become more and more pragmatic in my approach to lean manufacturing. I am not a purist when it comes to methodology. In fact, in this book I am sharing with you information based on my own personal research, true-life experiences, and lessons learned through the implementation of lean principles within a number of companies. It is this broad-based experience that has allowed me to develop such a pragmatic approach. My experience has taught me that, although a specific philosophy may work well with one particular project or company, it may not work as well universally across other operations.

The information, time frames, and methodologies contained within this book are geared primarily for operations that have 300 to 500 employees. The content was written for an audience operating at the level of plant manager, project manager, or manufacturing manager within a business, although most certainly schedulers, planners, industrial engineers, and first-line supervisors can also benefit from this material. The book provides tools and techniques that can be used for both high-volume/low-mix and low-volume/high-mix product environments. Although many of the techniques are designed for discrete unit manufacturing operations, those in the process industries could utilize many of the principles presented here, as well.

I realize that there are some of you who operate within an environment that does not require you to justify your position on lean manufacturing every step of the way and that such an environment will accept the need for lean management based on faith. This book was not written for you. This book was written for your colleagues — those who need to justify their position every step of the way and must constantly battle "hurdle management" to deploy their lean programs. You know who you are and you know what I am talking about. This book was written with you in mind.

Now, one does not learn how to be lean just from reading a book. It is through actual hands-on implementation that one learns what does and does not work in most situations. It is out on the shop floor where practical meets theory. It is in the actual work environment where one learns that to be successful it is more important to have a clear understanding of how these techniques work than a vague understanding of what the technique is.

This book has been divided into four parts. Part I provides an explanation of *why* a holistic approach to lean is so beneficial in securing sustained improvement; it provides an overall view of *what* to do. The purpose of Part II is to furnish the reader with an understanding of the concept of the Five Primary Elements; it explores in detail several aspects of each of the five elements. Part III was written in the form of a story to depict actual use of the techniques from the inception of a project to implementation in the factory; it helps the reader see *how* and *when* these principles are applied as part of a lean manufacturing program. Part IV presents case studies of six different companies that have taken on the challenge of changing their businesses and describes how the companies have deployed lean manufacturing within their facilities. Each case study was designed to reveal a different aspect of implementing lean manufacturing within an operation.

The entire book attempts to provide insight as to the choice and use of appropriate tools for assessment, analysis, design, and deployment of a successful lean manufacturing program. Although it does not cover every lean manufacturing aspect, issue, or situation, it does offer a road map that can guide a company toward the development of a lean manufacturing environment. Over the years, I have read about, witnessed, and heard of a great many implementations that have neither achieved their intended goals nor sustained results. My experience has led me to conclude that there are several reasons for the demise of these lean manufacturing programs: (1) no clarified expectation or vision as to what the new lean environment was to look like; (2) lack of a clear direction as to where to go and what to do next; (3) limited knowledge base for how to conduct the implementation; (4) significant focus on the mechanics of the new process but little attention paid to organization redesign issues connected with the change. These are key, critical issues that must be addressed for an implementation to be successful. The fact that many companies have neglected to do so has led me to write this book.

Acknowledgments

I would sincerely like to express my appreciation to all the people and multiple companies with whom I have had the opportunity to work over the years. I am extremely grateful to a great many of you for the tremendous knowledge I have received during the last 15 years. It is the vast diversity of your ideas and business situations that has allowed me to have the insight necessary to write this book. Even though hundreds of individuals have influenced the writing of this book, I would specifically like to thank: Phil Parry, of the Bourton Group, for his many hours of counsel and guidance during a time of tremendous upheaval in my life; Ron Aarns, of Boeing, for allowing me the opportunity to show what is possible no matter what the impending odds; John Paul, for giving me the chance to see an entirely new global world in a very short period of time; David Hall, Joe Costello, and Mike Bell, for their valuable insight, thoughts, and feedback both before and during the writing of this manuscript; Allie McArthy, for her technical support, advice, and editing; and most of all I want to thank my wife, Julie Ann, for staying the course with me over the past 15 years of learning, listening, and leading — without her consistent support, this book could not have been written.

Lean Manufacturing

Tools, Techniques, and How To Use Them

by

WILLIAM M. FELD

The St. Lucie Press/APICS Series on Resource Management

St. Lucie Press
Boca Raton • London
New York • Washington, D.C.

APICS®
THE EDUCATIONAL SOCIETY FOR RESOURCE MANAGEMENT
Alexandria, Virginia

Library of Congress Cataloging-in-Publication Data

Feld, William M.
Lean manufacturing : tools, techniques, and how to use them / by William M. Feld.
p. cm.
Includes bibliographical references (p.) and index.
ISBN 1-57444-297-X (alk. paper)
1. Production management. 2. Costs, industrial. 3. Production management — Case studies. I. Title.

TS155 .F4985 2000
658.5—dc21

00-059163

Direct all inquiries to St. Lucie Press, 2000 N.W. Corporate Blvd., Boca Raton, Florida 33431.

International Standard Book Number 1-57444-297-X
Library of Congress Card Number 00-059163
Printed in the United States of America 7 8 9 0
Printed on acid-free paper

The Author

William M. Feld is a client partner with Cambridge Management Consulting (CMC), a division of Cambridge Technology Partners. He has nearly 15 years of industry experience implementing lean manufacturing improvements and has conducted over 60 individual Kaizen events, implemented over 200 manufacturing cells, and managed several lean manufacturing programs (utilizing many of the concepts described in the book) for companies in a variety of industries. He has worked in the machining, sheet metal, industrial products, pneumatic tools, aerospace, electronics, power drives, and automotive industries.

Prior to his work with CMC, Bill was a manager of change management for Invensys, PLC, where he was responsible for the development and implementation of business change management programs for Invensys companies throughout North America. He has been a plant manager for Stanley Mechanic Tools and a manufacturing and materials management consultant for Ernst & Young, in addition to spending over 10 years in the aerospace and defense industry at Boeing (formerly McDonnell Douglas). Bill has also served as project manager for the implementation of several cellular manufacturing programs and has participated in the implementation of two MRP II systems. He has held multiple line-management positions in manufacturing and materials management. Bill received his Master of Business Administration degree in operations management, earned a Bachelor's degree in business administration, and is certified in production and inventory management (CPIM) with APICS. He can be contacted at (314) 442-9768 or william.feld@worldnet.att.net.

About APICS

APICS, The Educational Society for Resource Management, is an international, not-for-profit organization offering a full range of programs and materials focusing on individual and organizational education, standards of excellence, and integrated resource management topics. These resources, developed under the direction of integrated resource management experts, are available at local, regional, and national levels. Since 1957, hundreds of thousands of professionals have relied on APICS as a source for educational products and services.

- **APICS Certification Programs** — APICS offers two internationally recognized certification programs, Certified in Production and Inventory Management (CPIM) and Certified in Integrated Resource Management (CIRM), known around the world as standards of professional competence in business and manufacturing.
- ***APICS Educational Materials Catalog*** — This catalog contains books, courseware, proceedings, reprints, training materials, and videos developed by industry experts and available to members at a discount.
- ***APICS: The Performance Advantage*** — This monthly, four-color magazine addresses the educational and resource management needs of manufacturing professionals.
- ***APICS Business Outlook Index*** — Designed to take economic analysis a step beyond current surveys, the index is a monthly manufacturing-based survey report based on confidential production, sales, and inventory data from APICS-related companies.
- **Chapters** — APICS' more than 270 chapters provide leadership, learning, and networking opportunities at the local level.

- **Educational Opportunities** — Held around the country, APICS' International Conference and Exhibition, workshops, and symposia offer you numerous opportunities to learn from your peers and management experts.
- **Employment Referral Program** — A cost-effective way to reach a targeted network of resource management professionals, this program pairs qualified job candidates with interested companies.
- **SIGs** — These member groups develop specialized educational programs and resources for seven specific industry and interest areas.
- **Web Site** — The APICS Web site at http://www.apics.org enables you to explore the wide range of information available on APICS' membership, certification, and educational offerings.
- **Member Services** — Members enjoy a dedicated inquiry service, insurance, a retirement plan, and more.

For more information on APICS programs, services, or membership, call APICS Customer Service at (800) 444-2742 or (703) 237-8344 or visit http://www.apics.org on the World Wide Web.

DESCRIPTION OF LEAN MANUFACTURING

I

1 Lean Manufacturing: A "Holistic" View

What Is Meant by *Holistic*?

What is meant by the word *holistic*? Is it meant to imply a well-rounded perspective? Is it used to describe an overall state of wellness? Does it mean all-encompassing? If we check the definition according to Webster's English Dictionary, holistic means "emphasizing the organic or functional relation between parts and wholes." Now, none of these definitions of holistic is necessarily wrong; however, when associated with our description of lean manufacturing, the concept of holistic is meant to imply the interconnectivity and dependence among a set of five key elements. Each individual element is critical and necessary for the successful deployment of a lean manufacturing program, but no one element can stand alone and be expected to achieve the performance level of all five elements combined.

Each of these elements contains a set of lean principles which, when working together, all contribute to the development of a world-class manufacturing environment, often reflected by a company inventory-turn level of 50 or higher. As described by Schonberger in his book, *World Class Manufacturing: The Next Decade*, inventory turns provide comparable anecdotal evidence of the level of performance of a company regardless of changes in economic swings, monetary policies, trade practices, or internal company manipulations: "We need not rely on case studies or news clippings. One statistic extractable from corporate annual reports tells the story with surprising accuracy: inventory turnover (cost of sales divided by on-hand inventory).

It happens that when a company manages its processes poorly, waste in the form of inventory piles up."[17]

Not only are these lean principles interactive and co-dependent, but there is also a fundamental relationship that exists among these principles as to the sequence by which they should be deployed. So what exactly are these five elements and what makes them so co-dependent?

Description of the Five Primary Elements

The Five Primary Elements for lean manufacturing are (1) Manufacturing Flow, (2) Organization, (3) Process Control, (4) Metrics, and (5) Logistics (Figure 1.1). These elements represent the various facets required to support a solid lean manufacturing program, and it is the full deployment of these elements that will propel a company on a path toward becoming a world-class manufacturer.

Following is a basic definition of each of the Five Primary Elements:

- *Manufacturing Flow:* The aspect that addresses physical changes and design standards that are deployed as part of the cell.
- *Organization:* The aspect focusing on identification of people's roles/functions, training in new ways of working, and communication.
- *Process Control:* The aspect directed at monitoring, controlling, stabilizing, and pursuing ways to improve the process.
- *Metrics:* The aspect addressing visible, results-based performance measures; targeted improvement; and team rewards/recognition.
- *Logistics:* The aspect that provides definition for operating rules and mechanisms for planning and controlling the flow of material.

These primary elements provide full coverage of the range of issues that surface during a lean manufacturing implementation. Each element focuses on a particular area of emphasis and compartmentalizes the activities. Even though each element is important on its own for the deployment of a successful lean manufacturing program, the power comes from integration of the elements. For instance, Manufacturing Flow sets the foundation for change. People see activity on the shop floor, furniture being moved (sometimes for the first time), machines or floors or walls being painted, and areas being cleaned up. Excitement and energy surround this visible change. Add to this the less than visible changes in infrastructure relative to organizational roles and responsibility, new ways of working, training of personnel, multi-

Manufacturing Flow
1. Product/quantity assessment (product group)
2. Process mapping
3. Routing analysis (process, work, content, volume)
4. Takt calculations
5. Workload balancing
6. Kanban sizing
7. Cell layout
8. Standard work
9. One-piece flow

Process Control
1. Total productive maintenance
2. Poka-yoke
3. SMED
4. Graphical work instructions
5. Visual control
6. Continuous improvement
7. Line stop
8. SPC
9. 5S housekeeping

Metrics
1. On-time delivery
2. Process lead-time
3. Total cost
4. Quality yield
5. Inventory (turns)
6. Space utilization
7. Travel distance
8. Productivity

Organization
1. Product-focused, multi-disciplined team
2. Lean manager development
3. Touch labor cross-training skill matrix
4. Training (lean awareness, cell control, metrics, SPC, continuous improvement)
5. Communication plan
6. Roles and responsibility

Logistics
1. Forward plan
2. Mix-model manufacturing
3. Level loading
4. Workable work
5. Kanban pull signal
6. A,B,C parts handling
7. Service cell agreements
8. Customer/supplier alignment
9. Operational rules

Figure 1.1 Five Primary Elements of Lean Manufacturing

function teaming, and identification of customer/supplier relationships. Finally, add the visible presence of shopfloor measurements reflecting status, equipment being repaired, graphic work instructions being posted at work stations, and machine changeover times being recorded and improved. These primary elements complement one another and are all required to support each other as part of a successful implementation.

Most lean manufacturing initiatives focus on the primary elements of Manufacturing Flow, some on Process Control and areas of Logistics. Once in a while, there is the mention of Metrics and some discussion regarding Organization, usually training. This lack of attention to the whole is a shame, because it is the culture changes in Organization and the infrastructure improvements in Logistics that institutionalize the improvements and provide for sustained change within the organization. When initiatives focus on just the mechanics and techniques (indicative of both Manufacturing Flow and Process Control), the improvement is more about calculations and formulas than it is about improving workforce capability. Anyone can read a book, run a numbers analysis on demand behavior, calculate takt time, and establish a U-shaped layout, but doing so is not what will make a company differ from its competition. True competitive advantage comes from instilling capability within the workforce, and this can only be accomplished through: (1) achieving demonstrated knowledge transfer by building an empowered workforce, (2) engaging all employees within the business by steering their collective energies in the same direction, and (3) empowering the workforce with clarified expectations, common purpose, and accountability to get the job done. An organization with this capability can be neither copied nor bought by the competition; it must be designed, developed, directed, and supported.

This book focuses on the relationships among each of the primary elements and provides a "how-to" road map for implementing lasting change. In order for these primary elements to function properly, they must be implemented in the form of stages or "building blocks." Specific foundation prerequisites must be met prior to deployment of subsequent stages. The initial stages contain criteria that must be satisfied before implementing subsequent stages. These criteria are like the prerequisites for some college courses. The first-level activities must be completed to serve as building blocks for subsequent stages. It is imperative that these stages be followed to avoid jeopardizing the implementation and to assure success in deploying the lean manufacturing program as quickly as possible for maximum benefit. Part III of this book will identify those stages and explain the appropriate sequence for implementation.

Lean manufacturing, as described in this book, is primarily focused on designing a robust production operation that is responsive, flexible, predictable, and consistent. This creates a manufacturing operation that is focused on continuous improvement through a self-directed work force and driven by output-based measures aligned with customer performance criteria. It develops a workforce with the capability to utilize the lean tools and techniques necessary to satisfy world-class expectations now and into the future. As noted by Conner in *Managing at the Speed of Change:* "People can only change when they have the capacity to do so. Ability means having the necessary skills and knowing how to use them. Willingness is the motivation to apply those skills to a particular situation."[3] Viewing lean manufacturing from a holistic perspective should be able to satisfy the need to have both ability and willingness.

2 Lean Manufacturing Approach

The first step required on this journey toward creating a lean manufacturing environment is to recognize where we are currently. We must demonstrate an understanding as to why we need to change, and we must determine why it is important that we make a change. What are the business drivers that have caused this intrusion of lean manufacturing into our lives and why should we care to participate? Answers to these questions are required in order for people to become engaged in the change process. How we handle the responses to these questions is critical to our success. Motivation, tenacity, leadership, and direction all play key roles in the successful deployment of a lean program. If we as individuals are not motivated to go down this path, if we do not have a direction to guide our next steps, and we do not have the tenacity to stay with the journey when it becomes bumpy, we may as well not begin.

In order to understand the current situation, we may need to conduct a self-assessment that will provide a sounding board or reflective mirror for our operating condition as it stands today. It will provide feedback regarding where we currently demonstrate capability, and it will reveal gaps between how things are being done today and what are considered to be sound lean practices. To provide some level of insight into this gap, one need only to look at the landmark MIT study conducted by Womack, Jones, and Roos (see *The Machine That Changed the World*) to understand how far some operations are from being lean. Facilities that are considered lean operate with far fewer resources as compared to those facilities that operate as mass producers: "Lean production vs. mass production: 1/2 the human effort in the factory, 1/2 the manufacturing space, 1/2 the investment tools, 1/2 the engineering hours, 1/2 the time to develop new products."[26]

It is only when we are honest with ourselves as to where we are that progress can really begin to make significant change. Benchmarking against a defined criteria and determining our performance gap are ways to begin building a story line for why we need to change (see Figure 12.2).

It is this story line that must be communicated to the organization in order to win support for a change program. By the time company leaders come to the conclusion that they need to change the company, it is usually after several months or years of seeing profits shrink through revenue loss at the top line or market share erosion. Usually, they have been looking at the data and reviewing the numbers for quite some time. When they finally do come to the inevitable conclusion that change is necessary, these same leaders need to inform the entire organization as to the scope of what they are changing and why. One cannot deploy a major change such as lean manufacturing and expect it to endure without engaging the entire work force. If one does not present a compelling story as to why change is necessary, employees are not likely to become engaged with the program. This is not to say that those initiating the change will have all the answers at this initial phase (because they won't); however, they should be able to explain why it has become necessary to conduct business in a different manner.

After having gone through the self-assessment and reaching agreement that there is a need for change, the next step is to assemble a team to design, develop, and deploy the lean manufacturing program. There are some general guidelines to follow when selecting a team and formally launching a project. First, the team must be full time; part-time teams give part-time results. If this project is not serious enough to launch with full force, do not bother to begin. Part-time members are only partially dedicated, which means they have other priorities and are not completely focused on the task at hand. It is better to dedicate three people full time than to staff a team with 12 part-time resources. Part-time teams simply do not work.

Second, roles within the team and the way in which team members interact with one another are quite important. It is imperative that all members understand their roles on the team and why they were selected for the assignment. When assessing project team candidates, it is important to keep in mind selection criteria and to have an understanding of what attributes are required. The following would be a good starter list of desired attributes:

- Open minded
- Effective communicator

- Results oriented
- Self-confident
- Resilient to change
- Challenger of the *status quo*
- Group facilitator
- Trusted judgment
- Influential within the organization

In addition to each team member's experience and expertise, an individual's preference toward taking on a particular role is an important factor in the successful outcome of a team's ability to deliver a project. Meredith Belbin* has done a significant amount of research in this area and has concluded that team role preference can have a considerable impact as to whether a team will perform successfully or not. Utilization of his material can provide some valuable insight into the appropriate makeup of project teams.

After the team has been selected, they must be mobilized. To accomplish this, the team will need to generate two key documents: a project charter and project milestone plan. The charter defines the project's purpose, objectives, and outcomes. The milestone plan identifies major segments of the project, the time frame for completion, and a sequence of major events. The milestone plan should be based on a lean manufacturing road map (Figure 2.1), which provides a common understanding for the team as to specific phases of the project.

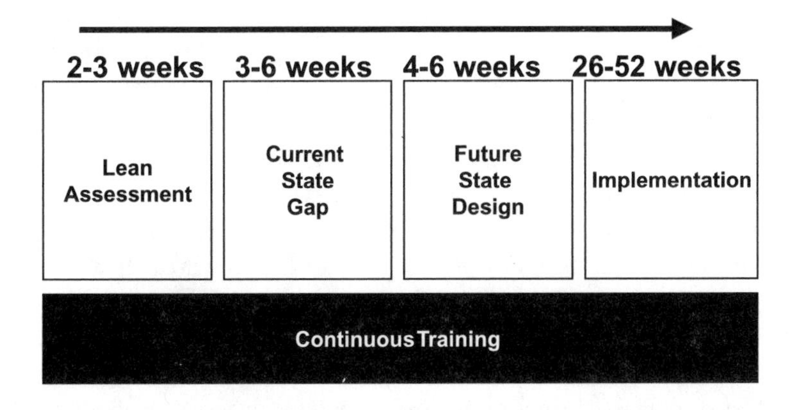

Figure 2.1 Lean Manufacturing Road Map

* Meredith Belbin is a British professor who has conducted nearly 30 years of research on teams, team dynamics, and developing insight into what makes successful teams work (see Belbin Associates' Website: www.belbin.com).

In addition to the project charter and milestone plan, the following elements should also be considered as necessary ingredients for the practice of good project management:

- Project protocol (team meeting time, place, duration, format)
- Project organization (steering committee, project owner, leader, etc.)
- Defined roles/responsibilities (for organization structure)
- Risk mitigation management (identifying and resolving potential risk)
- Hazard escalation management (rules for elevating problems)
- Project schedule (deliverables, ownership, dependency, resources)
- Issue log (catalog of project issues, action, dates, ownership)
- Project book (living and historical documents of the project)

The team-generated charter and milestone plans (see Figures 11.1 and 11.2) provide the first documented clarification of project expectations for executive management and the project team. These documents are to be agreed to and signed off on by all parties in order to minimize the risk of missed expectations down the road. It is at this time that an announcement should go out to the rest of the organization explaining what is about to take place in regard to the lean program. This communication should: (1) express the need for looking at doing business differently, (2) identify who makes up the project team, (3) reveal the project milestone plan, and (4) clarify for employees what this project means to them.

Once the project team has completed the initial debriefing with management, they are ready to begin detailing the lean project elements, which would include the project's deliverables (those very black-and-white, tangible pieces of evidence that provide proof that an activity is complete), the defined work content for each of the project deliverables with assigned ownership (responsibility, accountability, and authority, or RAA), the establishment of resource staffing requirements, and the team's agreement on project management protocol.

Once the team is up and operating, it is time to get down to business. For the team, this means working their way through each of the lean road map phases. The first phase, that of Lean Assessment (Figure 2.2), is used to determine how the operation stacks up area by area and product group by product group from a lean manufacturing perspective. In this phase, the team tries to understand where areas of opportunity and leverage points exist within the business. They begin building the story line for not only why the business needs to change but also where and how much. This assessment looks at process performance issues relating to the Five Primary Elements by identifying waste or "muda" opportunities that exist within the business.

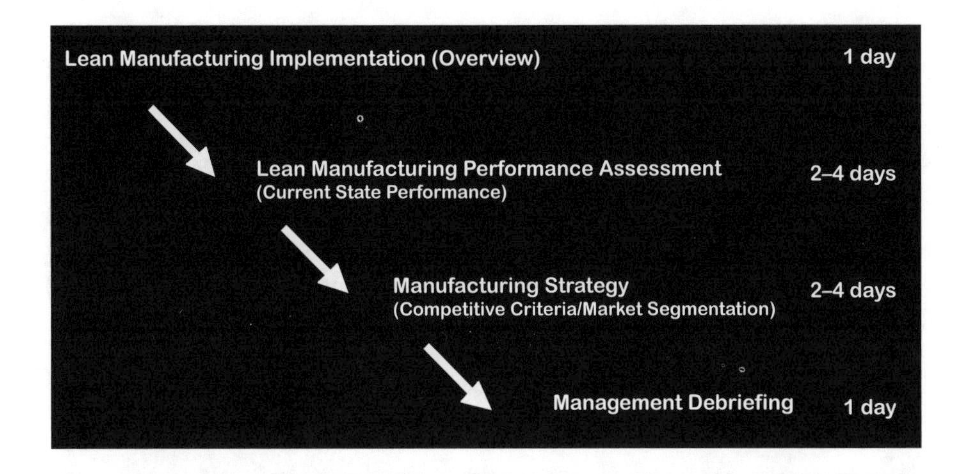

Figure 2.2 Phase 1: Lean Assessment

In addition to the internal search for opportunity, the outline of a manufacturing strategy is developed in order to assure alignment of the lean initiatives with the marketplace and to provide insight for the appropriate design criteria that are to be utilized in phase three, Future State Design. As Hunt clarified in *Process Mapping: How to Reengineer Your Business Processes*, it is necessary to understand the customer's performance expectations before designing a solution: "To simplify your product and process systems design, the process improvement team must first understand the customer's real requirements and priorities."[11]

This manufacturing strategy outline will identify which products compete in what markets and why. It also explores major competitors to understand the competitive criteria required for certain markets and determines where the team needs to leverage the change program to gain alignment with the current and desired customer base. Gunn emphasized this in his book, *Manufacturing for Competitive Advantage: Becoming a World Class Manufacturer*: "It is imperative to ascertain to the extent possible how effectively the competitors can manufacture products."[6] By aligning with marketplace requirements, the probability of leveraging bottom-line benefit for the business increases tremendously.

After Lean Assessment is complete, a second debriefing is conducted with executive management to report the findings and gain approval to move on to the next phase, that of documenting the Current State Gap (Figure 2.3). The Current State Gap provides the baseline measure of where the company is today. In this phase, the team:

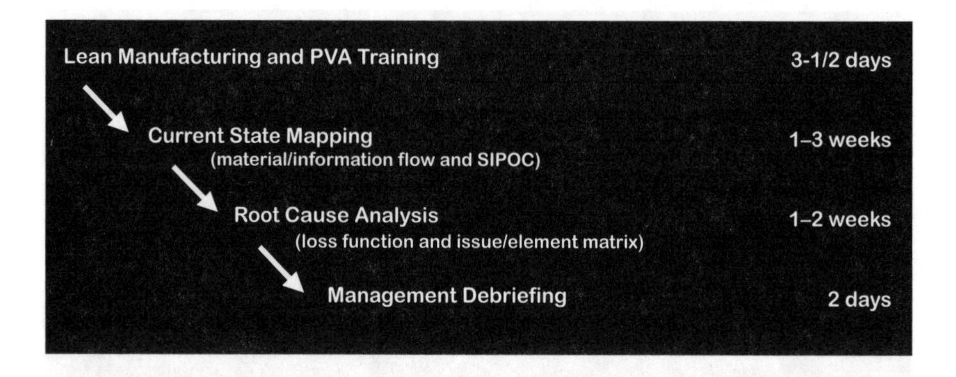

Figure 2.3 Phase 2: Current State Gap

- Receives training in process value analysis (PVA), lean manufacturing principles, and lean analysis tools
- Establishes process linkages through material and information flow mapping
- Quantifies where opportunities exist for waste elimination
- Generates design criteria based on the marketplace
- Creates a SIPOC (supplier-input-process-output-customer) map of all the major operational processes in order to understand customer/supplier relationships and required inputs and outputs that trigger these processes
- Analyzes current performance levels in regard to production loss function and waste elimination opportunities in order to prioritize implementation sequence and address risk
- Develops a "quick hit" list for short-term improvements and establishes a baseline for demonstrated improvement

If this last item is given approval by executive management, the short-term improvements will be deployed as part of the third phase. This would allow the company to begin realizing benefits quickly and to initiate self-funding of the change program. In addition, it allows people to see action and results right away.

After investing 3 to 6 weeks to gain an understanding of the current state and to confirm that understanding with the major process owners, a management debriefing is conducted to inform executive management as to what was discovered. Executive management approval allows rite of passage to the third phase, which is focused on the Future State Design (Figure 2.4). In the Future State Design phase, the project team puts together an overall concept

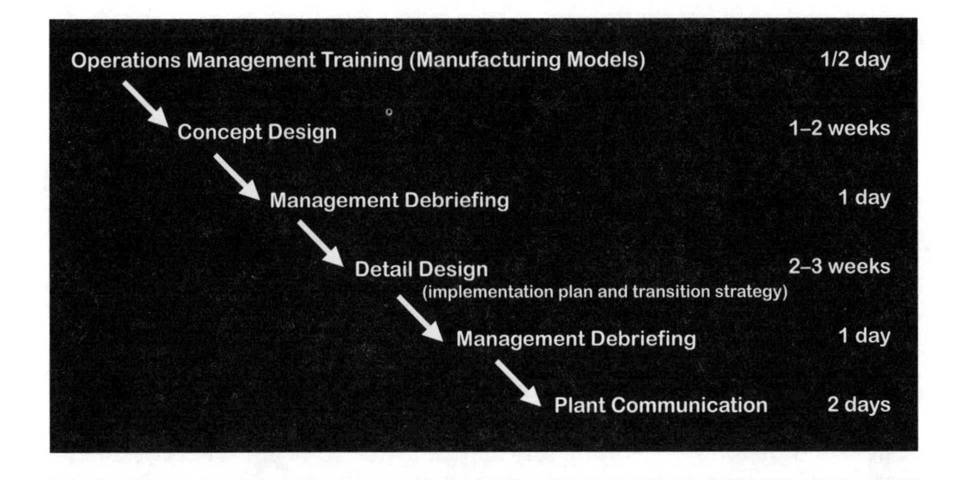

Figure 2.4 Phase 3: Future State Design

design of how the site should operate. This process will take approximately 2 to 3 weeks and includes:

- Determining what product groupings exist and how they would be produced
- Generating a general organization structure
- Producing a block layout for the plant
- Analyzing product demand behaviors and material/information flow
- Providing team training for the overall operations management structure (possibly including site visits to other lean operations) and exposure to different manufacturing architectures
- Confirming the concept design with major process owners
- Developing a new demand management process for logistics (order launch to product delivery)

The team's concept design is presented to executive management for review and approval. When blessed, the team focuses the next 3 to 4 weeks on the second half of phase three, the development of a detail design. The outcomes of this detail design include:

- Shopfloor staffing plans
- Cell workload analysis
- Transition strategy
- Implementation plan
- Defined exit criteria

- Shopfloor organization roles and responsibilities
- Confirmation of the detail design with major process owners
- Shopfloor training program
- Communication program

This package is presented to executive management for approval. Upon agreement as to the lean program rollout, a communication regarding the implementation plan should go to all employees explaining what was discovered, who was involved, what was decided, where the organization is going, and where all employees fit in. It is at this time that the project team moves into the fourth phase, Implementation.

Now that the team has spent the last 9 to 15 weeks on assessment, analysis, design, and planning, it is time for the real action to begin. The investment in time and resources spent up front to understand the current process and design the future state can now quickly payoff. It is through the definition of a design criteria, the description of marketplace and customer value opportunities, and the establishment of improvement initiatives around product groupings that alignment of the lean manufacturing program will leverage rapid benefits during deployment. This logic is similar to that described by Womack and Jones in *Lean Thinking: Banish Waste and Create Wealth in Your Organization*: "A firm might adopt the goals of converting the entire organization to continuous flow with all internal order management by means of a pull system. The projects required to do this might consist of: (1) reorganizing around product families, with product teams taking on many of the jobs of the traditional functions; (2) creating a 'lean function' to assemble the expertise to assist the product teams in the conversion; and (3) commencing a systemic set of improvement activities to convert batches and rework into continuous flow."[25]

The implementation of manufacturing cells is now conducted though a series of stages via "Kaizen events." These stages serve as building blocks and set the foundation for subsequent stages (Figures 2.5 and 2.6). For example, implementation of the first stage includes:

- Establishing the baseline cell design
- Balancing the cell to takt time
- Documenting the standard work content
- Establishing visual controls
- Creating the operating rules
- Introducing intra-cell material pull
- Defining team roles and responsibilities

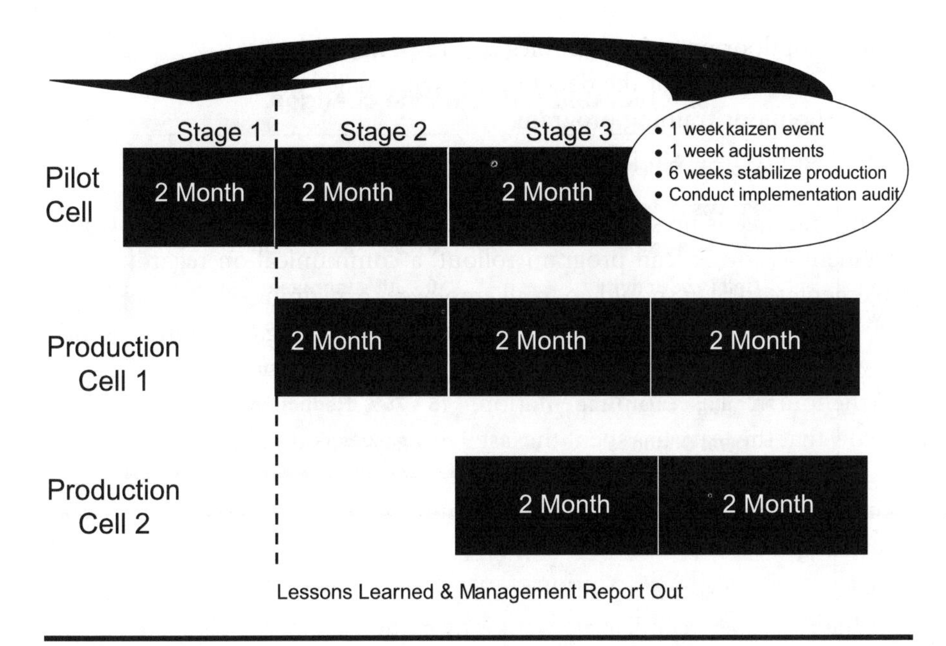

Figure 2.5 Implementation Methodology

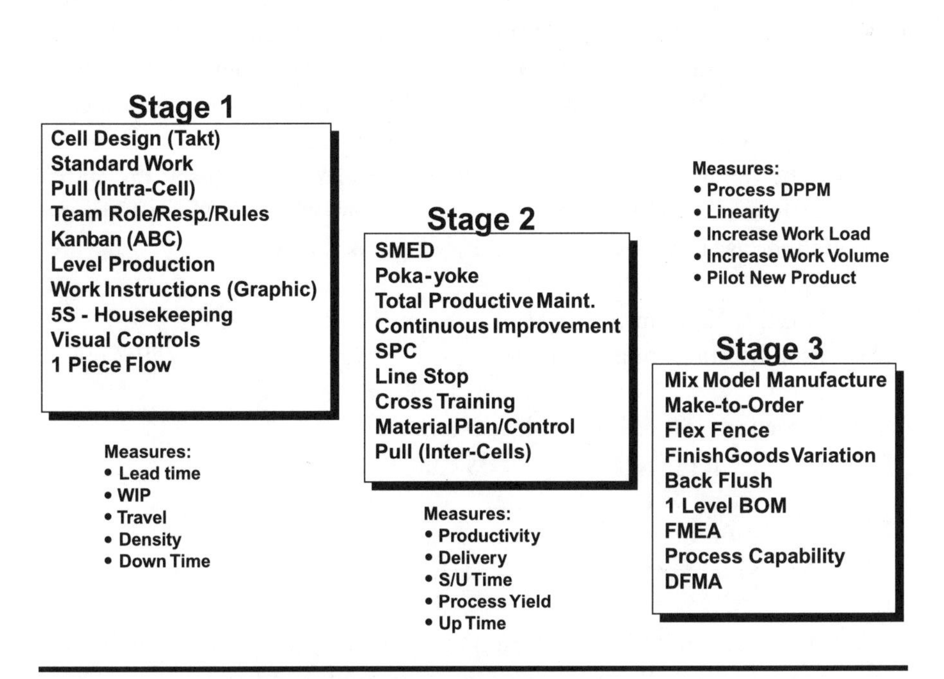

Figure 2.6 Lean Manufacturing Principles

Kaizen projects - resulting in the following demonstrated performance changes:

>Work in Process	30 - 90% Reduction
>Process Steps	25 - 75% Reduction
>Mfg. Lead time	20 - 90% Reduction
>Cell Productivity	0 - 30% Increase
>NVA Activities	25 - 50% Reduction
>Space Utilization	10 - 50% Reduction
>Change Over Time	15 - 75% Reduction
>Travel Distance	30 - 80% Reduction

Figure 2.7 Expected Benefits

When the one-week Kaizen event is over, the second week is spent tweaking the process and allowing for adjustments. This is due to the fact that not everything can be implemented in its final form during the first week. After about 6 weeks of operation, the process should be stabilized and performing at targeted performance levels. At this point, a lean manufacturing audit (see Figure 12.4) should be conducted to make sure the implementation is exhibiting lean manufacturing characteristics and has demonstrated a significant change in performance (Figure 2.7).

Once the cell is performing at the desired level and has passed the audit, the cell team is allowed to pursue the second stage, which is deployed in the same manner as stage one; however, this stage focuses on:

- Rapid utilization of single-minute exchange of dies (SMED)
- Establishment of a formal total productive maintenance (TPM) program
- Incorporation of Poka-yoke devices
- Utilization of statistical process control (SPC)
- Team member cross-training
- Utilization of continuous improvement tools
- Deployment of inter-cell pull system

Again, there is a 6- to 7-week period for stabilization to ensure that desired performance levels are being achieved and to conduct a formal audit. Once the second stage is completed, the cell team qualifies for advancement to the third and final stage, which is really where world-class performance capabilities are achieved through the cell's capability to:

- Perform mix-model manufacturing
- Deliver make-to-order production
- Convert to a one-level bill of materials (BOM)
- Take advantage of finished-goods variation techniques
- Support flex-fence demand management
- Utilize material backflushing
- Conduct a failure mode and effects analysis (FMEA)
- Calculate process capability (CpK)
- Contribute to the assessment of products through design for manufacturing/assembly (DFMA) principles

For purposes of risk mitigation, the first cell needs to be deployed as a pilot cell, where over 50% of all lessons learned are obtained. Capturing those lessons learned and utilizing them during the deployment of subsequent production cells is invaluable. As each cell is implemented and becomes self-sustaining, look to link individual production cells together through customer/supplier alignment with inter-cell Kanbans. It is important to make sure that individual cells are stable before interconnecting them with other cells. If they are not, the internal supply chain is put at risk.

Once 50% of the production cells are in stage two and well on their way toward self-sustaining implementation, it is time to take the focus of the project team off the shop floor and to begin to pursue improvements in other areas of the business. This is in keeping with the advice given by Imai in *Gemba Kaizen*: "Gemba Kaizen becomes the starting point for highlighting inadequacies in other supporting departments and identifies systems and procedures that need to be improved."[12] The first area to address, therefore, would be that of customer interface for order processing and demand management. By this time in the project, enough improvement has been demonstrated on the shop floor that it is time for the team to work its way down the value stream toward the customer base.

The second area of focus would be that of product development. Now that the shop floor has a greater understanding of its capability, they can deliver extremely valuable insight into product designs and also contribute to the new product development process. The third area of focus would be redesign of the organization from where it is now to something that is more reflective of the new manufacturing architecture, where form would begin to follow function. The fourth area would be that of the external supply base.

- The project will be given the time necessary to deploy.
- The project will be given resources (funds and people).
- The project will be given a full-time/focused team.
- The project will be given clear expectations.
- The project will have an identified management sponsor.
- The project will have access to management guidance.

Figure 2.8 Project Management Assumptions

Now that a solid working model exists inside the factory and confidence has been gained in using the lean tools and techniques, it would be appropriate for the project team to work up the value stream toward the supplier base.

Even though these initiatives are listed in a serial manner, they can be addressed in parallel; however, that is only recommended with a word of caution. A company has only so many resources and realistically cannot address more than three to five company-wide initiatives at any one time. In addition, if lean manufacturing cannot be demonstrated at your own facility, it would not be wise to expect a customer or supplier to jump on board unless they have already been conducting lean manufacturing initiatives within their facilities. Some activities can be done in parallel, but be aware of capability surrounding the entire supply chain. Remember that a chain is only as strong as its weakest link.

Ingersoll Engineers, in *Making Manufacturing Cells Work,* probably best summarized this overall approach to lean manufacturing: "The greatest benefits are realized quickly in companies that include all affected functions from the beginning of the flexible manufacturing cell (FMC) project. …Cells simply don't work well, if at all, when they are not part of an overall strategy of change undertaken by their users. Cells standing alone are worthless. They are isolated islands remote from the rest of the world."[13] For any project team to be successful, a number of project management assumptions are required (Figure 2.8).

The one final question that remains for management to address to ensure a successful conclusion to the lean manufacturing program is "Are you willing to do what it takes to become a world-class manufacturing organization?" (Figure 2.9). If management is not willing to commit to these issues, then it is not recommend that they pursue deployment of a lean manufacturing program. If these key ingredients are not present within the spirit of the operation, the improvement initiative will struggle severely and often time result in failure.

- Can you impact production for 1 week? 2 weeks? 3 weeks?
- Can you dedicate 3 to 8 people for 6 to 9 months?
- Can you endure failure and mistakes before success and improved performance are fully realized?
- Can you provide commitment, even when you do not see major results after 2 months?
- Can you hold the course for 18 to 24 months?

Figure 2.9 "Are You Willing To...?"

Part I of this book has provided insight into the overall aspects of a holistic lean manufacturing program and has demonstrated how to set up and manage a lean program. Part II will describe in greater detail each aspect of the Five Primary Elements of lean manufacturing.

FIVE PRIMARY ELEMENTS

3 Organization Element

Most project managers recognize that culture is one of the toughest things to change in any company. By definition, a company's culture is "those activities that go on within a company when management is absent." A company's culture contributes significantly in the formation of an organization's behavior and can be difficult to alter. Behaviors that relate specifically to a company's informal operating system have usually been cultivated over many years and may not support or align with new continuous improvement initiatives. A lean manufacturing implementation cannot survive within an old culture that does not support a new operating environment.

Many questions are asked by management and employees alike when facing a lean manufacturing implementation with its newly developed responsibilities. Who has ownership for products? What happens when a product leaves the cell? Is our touch labor workforce cross-trained sufficiently to operate in a lean environment? What does a cell mean to our company? Do we involve the union? Do we already have cell leaders, or should they be interviewed and selected? Who reports to the cell leader? What is the role of a cell leader? Is it just touch labor? Is it production control? Is it production engineers? Is it quality inspection?

All of the above are excellent questions and are usually overlooked when a lean manufacturing implementation is limited to equipment rearrangement and shopfloor layouts. There are over a dozen different cultural issues involved with these questions, and any one of them can stop an implementation dead in its tracks. Most factories today still require human resources; therefore,

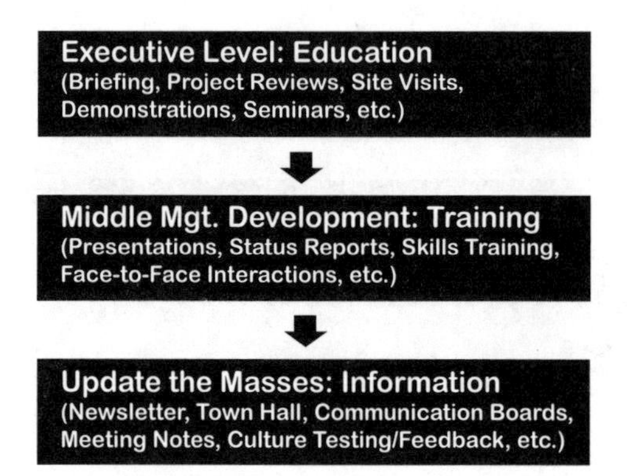

Figure 3.1 Communication Planning Hierarchy

people play an instrumental role in the success or failure of factory improvement initiatives. Many initiatives have failed due to the neglect of these cultural issues. So how does one handle this influx of cultural-related questions? What methods are utilized to tackle these issues? To address these questions in a logical manner, individual areas have been identified for discussion here:

1. Communication Planning
2. Product-Focused Responsibility
3. Leadership Development
4. Operational Roles and Responsibilities
5. Workforce Preparation

Communication Planning

"What's in it for me and where do I fit in?" If you want to get people's attention, nothing piques their interest more than threatening their jobs or changing the way in which they do their work. Do not keep them in the dark about the proposed changes. Fear is the human emotion that keeps us alert during times of duress and keeps us alive in situations of great danger. Fear is a motivator. When individuals are threatened by actions that have the potential to impact their livelihood, they protect and try to preserve those things over which they have control and fend off those over which they do not. Thus, it is best not to generate fear of an initiative before it even gets off the ground. Develop a communication plan that is focused at three levels within the organization, and tailor the content and subject matter to each (Figure 3.1). Utilize various forms of media to distribute the message and provide a clear understanding

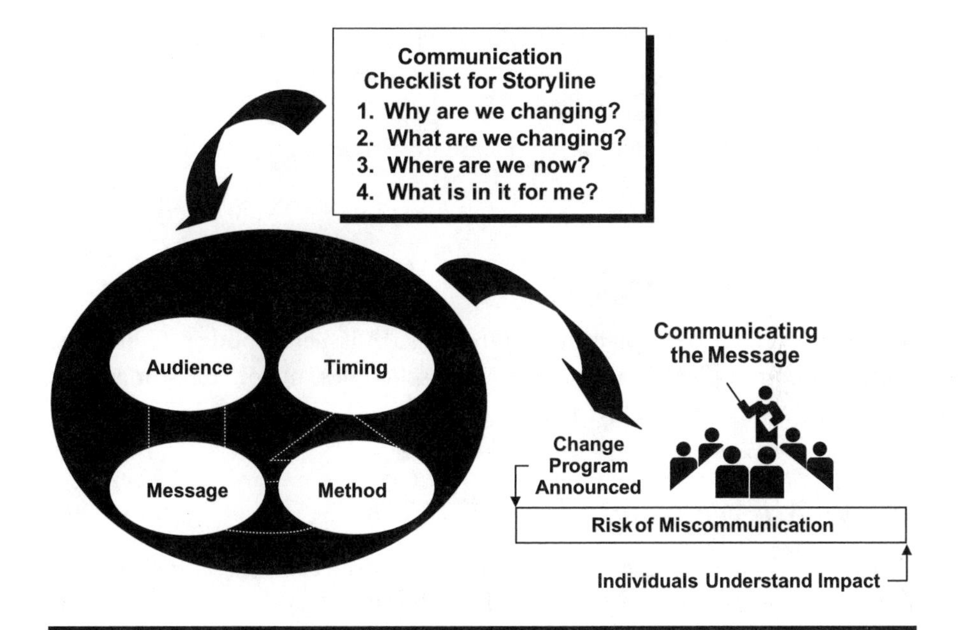

Figure 3.2 Aspects of Communication Planning

about what is required for each audience. Executive management requires understanding and the ability to approve. Middle management needs a significant amount of education and training. The masses require validation and assurances that they are included in the project's deployment.

When presenting the plan to the different levels within the organization, make sure the following four questions are answered as a part of the communications (Figure 3.2):

1. Why are we changing?
2. What are we changing?
3. Where are we now?
4. What's in it for me?

Why Are We Changing?

Put together a presentation that is applicable to all employees in the organization. It should be a relatively high-level briefing that:

1. Describes why the business is making a change in this direction (business environment, competitive position, market opportunity, etc.)

2. Explains how various employees will fit into the new environment (who could be affected, levels of management, potential role changes, etc.)
3. Clarifies operational expectations (e.g., 35% improvement in operational performance, 22% increase in market share, 18% reduction in total costs)

What Are We Changing?

Show an overall project plan that addresses such issues as budget (where the money is coming from), implementation schedules, major milestones, and areas involved in the deployment.

Where Are We Now?

Describe the stage of the game at which the project is currently residing. Publish a regular newsletter or e-mail for the shop floor and office environment to keep people up to date with how the implementation is progressing.

What's In It for Me?

Address the following issues from the perspective of the individual employee:

1. Where do I fit into the new organization?
2. How will this change affect the way I perform my job?
3. How do I benefit or value from this change?

Addressing these four questions will begin to engage people with the change process and help secure their involvement over the long haul.

Product-Focused Responsibility

Think about your own organization for a minute and ask yourself this question: If a customer called today and asked who in your organization was responsible for the quality and delivery of product XX-1324, what would your answer be? If there is a delivery problem, do we turn to production control? If there is a quality problem, do we turn to inspection? If there is a product cost issue, do we inquire with accounting? Using such logic to solve these problems can be attributed to the ever-popular "functional organization." Everyone has a piece of the action but no one has responsibility for the whole, except perhaps at the plant manager level. Ask yourself, is the plant manager the appropriate person to be addressing specific questions about products? Shouldn't the people with assigned responsibility for the product be answering questions about the product? The answer is yes. Yes, they should!

According to Schonberger in *World Class Manufacturing: The Lessons of Simplicity Applied*, this point is extremely important for reducing infighting and waste in the process: "World Class Manufacturing (WCM) requires organizing for quick flow and tight process-to-process and person-to-person linkages. The overriding goal is to create responsibility centers where none existed before. When responsibility centers are operating, the procrastinating, finger-pointing, and alibiing fade; the stage is set for conversion to a culture of continuous improvement."[16] How, then, does one bring about this realignment of ownership?

This realignment can be achieved by addressing three aspects of lean manufacturing:

1. Developing a responsive material and information flow infrastructure (Logistics; see Chapter 5)
2. Designing a flexible manufacturing architecture (Manufacturing Flow; see Chapter 6)
3. Transitioning ownership through the concept of empowerment

All of these aspects were referred to by Mahoney in *High-Mix Low-Volume Manufacturing*: "Employee participation and empowerment are results of the production situation. Attempts to empower the workforce and obtain continuous quality improvement without a sound underlying system of support are doomed to failure."[14] The overriding premise here is that the responsibility for decisions and accountability for performance are delegated to employees in a given cell when they have the appropriate level of training, tools, and techniques by which to embrace this new ownership.

A critical change required to support this realignment of ownership is to make it clear that cells are formed around products and products are the responsibility of cell members under the direction of a cell leader. The cell team has responsibility, accountability, and authority (RAA) for product quality, delivery, cost, and any other element or aspect of that product that is assigned to the cell level. Each cell should be provided with the resources

- A clear objective
- Need for intense concentration
- Lack of interruptions
- Clear and immediate feedback
- Sense of challenge
- Skills adequate to perform the job

Figure 3.3 To Be Successful, People Want...

necessary to carry out this mission. That does not mean that every organization's cells will look or be staffed in exactly the same manner, but it does mean that each company will assign the appropriate cell resources to match their given RAA.

An organization's size, level of manufacturing process complexity, level of cultural maturity in terms of empowerment, etc. are all factors in determining the makeup of cell organization structures. At a minimum, the cell should be staffed with a dedicated cell leader (who could manage more than one cell), identified touch labor personnel, and any required support resources (e.g., production control, production engineers, quality personnel, maintenance) necessary to carry out the mission of the cell team. It may be fiscally prudent to dedicate support personnel to more than one cell; however, each organization will have to determine a best fit for their own operation. Some organizations have established a two-tier structure in which the day-to-day activities (those occurring within 1 to 30 days) are handled at the shopfloor level and the month-to-month activities (those within 60 to 90 days) are managed at a level above the shop floor. This division of labor allows for the separation of resources for planning and execution. Resources above the shop floor can concentrate on preplanning and problem prevention without being consumed with firefighting taking place on the shop floor. The dedicated shopfloor resources can focus their energies on the product and executing day-to-day requirements.

In order for individuals and teams to be successful in an empowered environment, a few ingredients are required (Figure 3.3). If people are given clear expectations, the proper environment in which to concentrate, minimal interruptions, immediate and direct feedback, challenging goals, and the skills necessary to perform their jobs, positive performance results will be generated. When management creates this environment and nurtures these conditions, empowered, self-directed teams can flourish.

Leadership Development

Ask yourself, "Are the shop foremen I have running my shop floor today the leaders I want operating cells within my lean manufacturing environment of the future?" This is a very difficult question for many plant managers to answer because they have to determine whether the "down in the trenches" frontline supervisors who have gotten the organization where it is today are qualified to take it to the next higher level of performance for tomorrow. When an implementation considers only the physical aspects of a lean manufacturing project, this idea of proper leadership is never addressed. In order for a workforce to be truly empowered, it must first be equipped with the appropriate management skills and knowledge that will enable it to set its direction, maintain control over its destiny, and sustain continuous improvement after the initial implementation team is long gone. This does not happen by "teaching an old dog new tricks." This is not to say that shop foremen are not capable of leading and managing cells; however, a company that is transforming to a lean environment is establishing new mini-businesses, not new factory departments. We are not "rearranging the deck chairs" in this new environment. We are looking for leaders who can plan activities, set objectives, manage more than just task-based work assignments, and recognize cause-and-effect relationships relative to product cost. These are not positions to be filled by individuals who have been promoted up the ranks because of excellent shop knowledge. These are business managers who could very likely be required to interface with outside customers and suppliers. As stated by Tobin in *Re-Educating the Corporation: Foundations for the Learning Organization*: "Organizations are becoming flatter, with fewer levels separating the top officers of the company from the lowest levels. ...Work teams, whether within a single function or cross-functional, are becoming key organizational units. They are being given more and more responsibilities that used to belong to higher level managers — from problem solving to hiring to making capital investments."[24] Viewing the situation in this light, who do you want your next cell/business unit leaders to be?

How do you find these future leaders? Many of them currently work in the factory or at least within the company today. Consider, the next time you are in a meeting that includes employees from various functions across the business, who is exhibiting the following characteristics or management skills: planning, leadership, problem-solving ability, team building, technical competency, and interpersonal communication. These are the people you are looking for to fill leader roles. These are the people who will challenge the *status quo*. These are the people who will work with their direct reports to

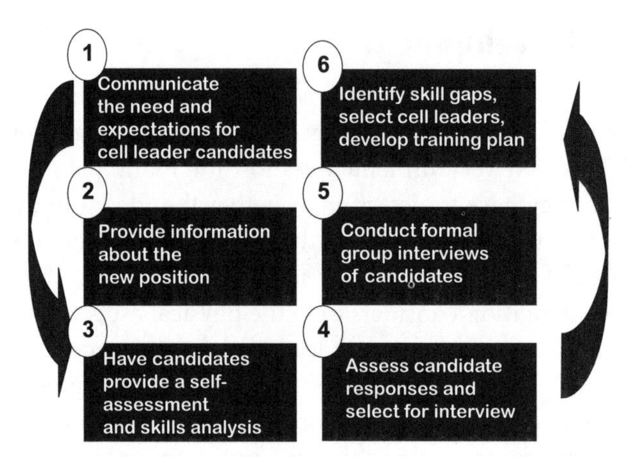

Figure 3.4 Cell Leader Selection Steps

accomplish a set of given objectives. However, if you cannot readily see and identify such personnel, do not despair; there is still hope.

By developing and deploying a formal selection and assessment process, a company can utilize a structured framework by which to select future cell leaders (Figure 3.4). It is highly recommended that some form of a formal process be used in the selection of cell leaders for three reasons: (1) the human resources department should be able to keep you out of hot water on the numerous legal issues surrounding employee discrimination; (2) you and the new cell leaders will be able to identify a training plan for those skills that are required for the position yet are lacking at the time of selection; and, most importantly, (3) your new leaders will be selected out of a field of their peers. They have been singled out as the "best" to fill this new position and will now directly be helping the company succeed with this new direction. What could be more rewarding for a self-motivated individual who has the desire to lead than to have his leadership qualities recognized through a formal assessment process and to be selected to manage a cell? When this highly motivated person, with leadership and team-building skills, is empowered to organize his team and set a course for continuous improvement, there will be no limit to what this team can accomplish.

Operational Roles and Responsibilities

Now that we have a cell leader and have assigned team members, we should be ready to move on to the next cell, right? Wrong! Do you think that within this new working environment the traditional functional roles will remain unchanged and unaffected? The roles and responsibilities of both touch labor and support personnel will be altered. Some positions will be changed more than others; nonetheless, they all will be different. The cell teams should be staffed with the minimum, flexible resources necessary for them to meet all operational performance objectives. This will almost certainly vary from cell to cell and company to company, but the fact remains that we are all doing more with less in this increasingly competitive global world of manufacturing. In light of this, the number one competitive weapon that comes to mind is flexibility; therefore, plan on staffing the cell with at least the minimum it needs to survive and allow continuous improvement to become a motivator. If the cell is developed with an overstaffed design, then when improvements are generated people will immediately need to be removed from the cell to address productivity objectives.

When operating within a union environment, be sure to include local union management participation during these design efforts. There will be issues that arise when operating in a union environment that do not arise in a non-union environment. The key to implementing lean manufacturing in a union environment is open and direct communication. How well the need for change and defining "what's in it for me" are communicated to the organization at the launch of the project can go a long way toward reducing conflict at this juncture. Numerous issues will arise when dealing with contract labor, such as flexibility across labor classifications, a limit to "individual job" contract language, method of pay, years of seniority, bumping rights, overtime allocation rules, etc. It is not that lean manufacturing cannot be implemented in a union operation (see case studies); it just requires additional considerations. Conflict resolution through joint problem-solving is critical to overcoming union and company management issues. Limiting disagreements to the facts and not opinions, agreeing on the direction to be taken and performance levels the company needs to achieve to survive and grow, and joint problem-solving to achieve those business results can significantly influence how far a lean manufacturing implementation will go and how quickly.

The first step in determining the roles and responsibilities of a cell team is to establish an agreed-upon focus (i.e., mission or charter) for the entire team on which they will concur and can channel their collective energies. This will allow the team to determine the functions required to make the cell work. The second step is to assign which functions the cell team members should and should not do. This is achieved by mapping out the operation, assigning responsibilities, and identifying the gaps. The third step is the development of an operational description or functional specification that defines the required tasks and responsibilities.

Once the functions required to operate the cell have been agreed upon, team members for each individual job function can write specific roles. This will not only help to eliminate the gray areas of functional responsibility, but it will also clarify for the human resources department what the new job descriptions are so they can utilize this documentation to sort out different pay grades and title changes. In addition, by involving union representation up front during the development of roles, they are cognizant of the changes and can highlight union contract issues early in the process. They retain ownership for the final product and can more easily mitigate concerns that may arise with the local union management.

Workforce Preparation

Although we introduced flexibility early in our discussion of operational roles and responsibility, this is where its impact can be felt on a minute-by-minute, hour-by-hour basis. Increasing the speed of workflow through the cell is one of the primary objectives for lean manufacturing; therefore, those individuals who actually touch the product (shape it, mold it, machine it, assemble it, etc.) are truly the only value-adding activity from the customer's perspective and need to be effectively deployed when producing the product. This means each touch labor employee ultimately will need to be capable of operating every process within the cell. This is more easily said than done, but the transformation has to begin somewhere. A recommended approach to initiating this transformation is to build a skills matrix (Figure 3.5), in which the people in the cell are listed on the *y*-axis and the processes or operations to be performed are listed across the top on the *x*-axis.

Filling out this matrix gives the cell team and cell leaders the means to identify areas and people requiring training. A recommended approach to soliciting input (because this can be a very uncomfortable part of the cell development process) is for the cell leader to ask his touch labor personnel: (1) what they can do well, and (2) what they cannot do because of any limitations (e.g., union contract, physical conditions). Do not ask them what they cannot do. This is a negative approach and puts the employee on the defensive. It will become evident soon enough as to what they cannot do when they have to begin performing at multiple workstations. By incorporating a validation process to clarify what is expected of the job, and validating performance in regard to those clarified expectations, the cell leader will be

Depth	3	4	3	3	4	4
Names	Table Saw	Gang Punch	Pin Route	Joggle	Drill Press	Testing
Sam	Can train others		Can train others	In training	Fully trained	Restricted
Joseph			Fully trained		Fully trained	In training
Gary	Fully trained			Can train others		In training
Gregg		In training			In training	Restricted
John	Fully trained	Fully trained			Can train others	
Leo	Restricted		In training	Restricted	In training	In training
Tom		In training		Fully trained		Can train others
Carol	Restricted	Can train others	Fully trained	Fully trained		Fully trained

Restricted · In training · Fully trained · Can train others

Figure 3.5 Cross-Training Matrix

able to develop a more accurate picture of the capability of the cell. Companies should develop a fair and unbiased validation process. In doing so, they may be able to take advantage of a skill-based pay scenario down the road. In addition to an inquiry as to what they can do well, ask the employee to rank their skills from strongest to weakest. This will help establish training plan priorities. After the matrix is complete, you should have a pretty good idea about what areas for improvement need to be addressed in the short term.

As cell team members become familiar with their new responsibilities, accountability for performance can begin to be established. Validation of actual performance and the use of control mechanisms that look at variation from plan both support adherence to standards and drive continuous improvement in the process. It is through this monitoring of the process, that we can keep our operational output performance in check, as we will see in the next chapter.

4 Metrics Element

In comparison with the other four elements (Organization, Logistics, Manufacturing Flow, and Process Control), Metrics (measurement) is the element that provides the primary focus for changing behavior. It is this element that ensures alignment between cell-level shopfloor activity and higher level company business objectives. It is this connection that is necessary for lean manufacturing improvements to appear on the bottom line. This chapter will describe how empowerment at the cell level to achieve operational objectives leads to improved performance, resulting in an impact on the company's bottom line. The metrics described will not be new, but they may be applied and managed in a manner that could be contrary to what some companies are accustomed.

No matter what company or what industry, we all have our fair share of metrics. There are metrics on cycle time, defects per unit, items shipped on schedule, direct labor cost, return on net assets (RONA), overtime, percentage of work orders released on time, cost of quality, hours of rework, cash flow, inventory turnover, etc. No company is lacking for reported measurements of performance. It is recognized that companies are spending valuable resources collecting, sorting, analyzing, and displaying these performance data and reporting them on a monthly, weekly, daily, and sometimes even hourly basis. If we, as companies, are spending this much effort on measurement why aren't all of our organization performing at "best-in-class" levels? Why are some of our organizations leading the pack while others are falling behind and some way behind? According to Hays, Wheelwright, and Clark (*Dynamic Manufacturing: Creating the Learning Organization*), it could very well be a matter of too much data and not enough information: "Measurements can provide useful information to managers who are trying to identify the sources of their problems or the reasons for their success. But most measurement systems in place today do not provide the kind of information needed by companies that seek to create a competitive advantage through manufacturing."[8] This chapter will explore some measurement formats that will enlighten us about this situation and will explain why an understanding of human behavior plays as much of a vital role in the success of performance improvement as the metrics themselves.

We are what we measure. …Improvement comes only from that which is visible. …A hidden problem reveals nothing. Although these statements have an element of truth to them, the real power of measurement comes from an individual's understanding of the measurement itself. The real trick to improving performance comes from an individual's definition, development, control, and understanding of cause and effect as they are related to the metric. Metrics that are developed by an outside entity and forced on a cell team are not likely to produce desired results. Metrics defined and developed by a cell team have a higher likelihood of resulting in a positive correlation between activity on the shop floor and desired performance. Understanding of the measurement, ownership of its results, and control over the factors that make it rise or fall are all important features necessary for the successful deployment of a measurement system.

This area of measurement will be looked at from several different perspectives:

1. DuPont model (a company view)
2. Output-based measures (a cell team's results)
3. Process-driven measures (infinite continuous improvement)
4. Goal alignment through policy deployment
5. Measurement definition and understanding (power to the people)

DuPont Model: A Company View

The DuPont model (Figure 4.1), which was developed by a French engineer in the 1940s, is an excellent tool to use to generate a "what if" analysis utilizing a company's income statement and the balance sheet. These time-honored instruments of the financial community are pivotal documents for reflecting the overall health of a company. It is through the intersection of these documents that the DuPont model becomes valuable as a performance measure.

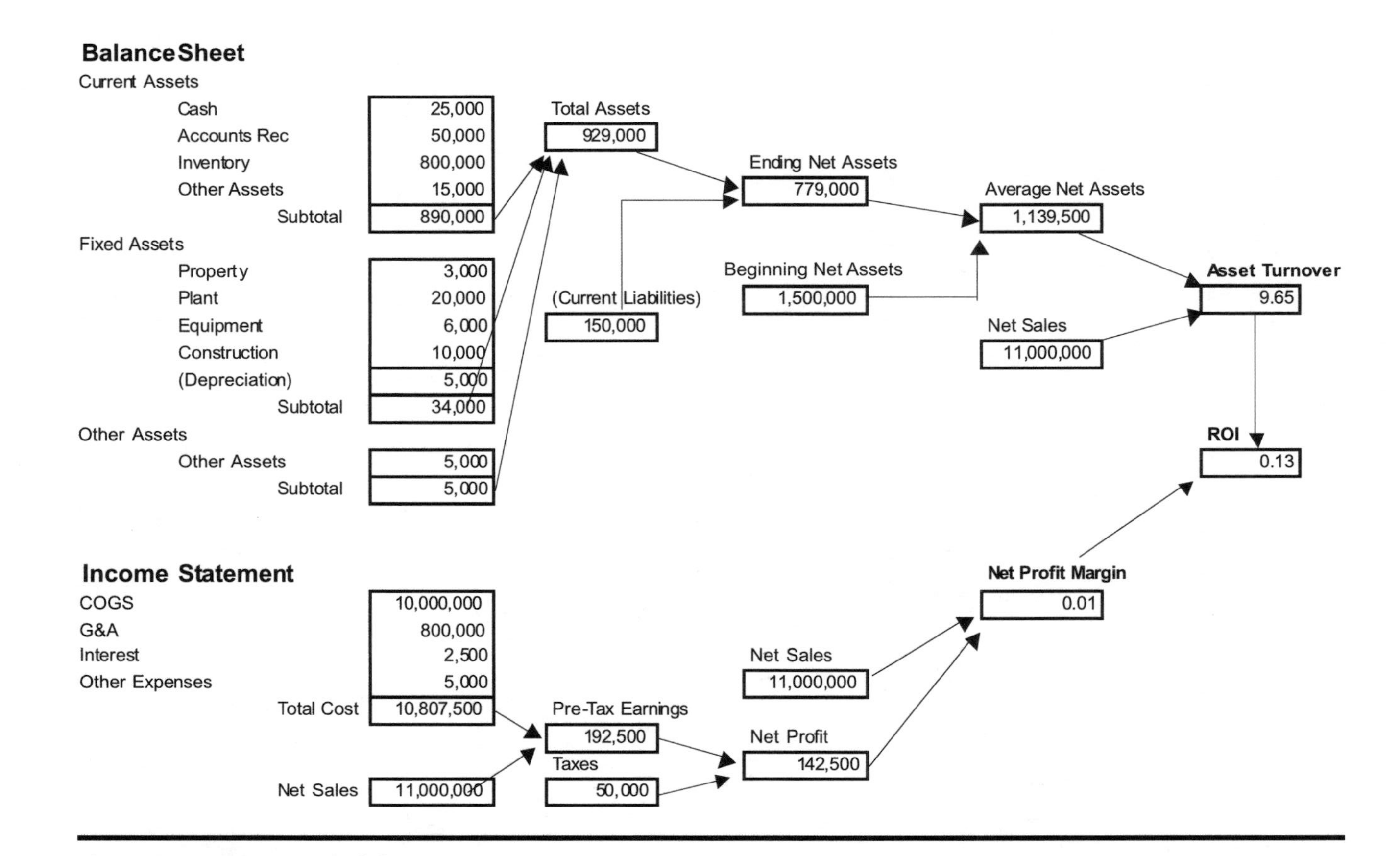

Figure 4.1 DuPont Model

By building relationships between particular line items it is possible to produce several different ratios as indicators of performance trends. In addition to monitoring trends, these ratios can serve as points of reference for industry comparisons, as well as a guide for establishing operational goals. Achievement of these goals can be played out through a "what if" scenario to determine whether improvement through a change in sales, an increase in asset turnover, or lower inventory levels will support operational objectives.

For example, if the plant can reduce inventory by 21%, then total assets will decrease and the asset turnover ratio will increase by 9%. With all other elements remaining equal, this will improve the return on investment (ROI) calculation by 5%. A second example would be if the cost of goods sold (COGS) is reduced by 7%, then total costs will decrease and net profit will improve by 4%. With all other elements remaining equal, this will improve the profit margin calculation by 2%.

This performance measurement method is useful at the top level within an organization to establish overall goals and objectives within the business. It is not designed for use at the cell level, where financial measurements tend to be less tangible. At the shop floor, in most cases it is better to utilize more tangible, physical measures of performance.

Output-Based Measurements: A Cell Team's Results

Typically, the only existing evidence of measures on the shop floor are measures based on performance for a specific individual employee or piece of equipment (e.g., how many hours were put in yesterday by Fred, or how many parts came off machine number 435 last shift, or what the yield of the drilling process was last hour). These are all measures of how a particular step in the process is performing, and they focus on what is called *localized optimization*. Companies monitor individual operations in the manufacturing process and assign accountability and take corrective action based on them. The problem with this type of measurement is that it: (1) drives the wrong behavior for continuous improvement, (2) does not really have product accountability focused on the customer, and (3) rewards optimization of the individual operation by sacrificing performance of the process as a whole. Again referring to Hays, Wheelwright, and Clark (*Dynamic Manufacturing: Creating the Learning Organization*), on the subject of product-focused vs. process-focused operations: "A product focus, on the other hand, is generally easier to manage because of its smaller size and total responsibility for a particular product or customer. This usually results in shorter cycle times, faster response to market changes, less inventory, lower logistics costs, and, of course, lower overhead."[8]

Following this theme of being product focused, a more effective measurement system would be one that establishes output-based measurements for the cell team around tangible products that go to a customer. These types of measures provide feedback on the performance of the overall process relative to the customer. Output-based measures assign accountability for all the operations contained within the manufacturing process. Responsibility is "cradle to grave" for the product. A focus on output measures drives continuous improvement in that someone is accountable to an end customer for the performance of a product and has the responsibility to correct any problems encountered by that customer.

Two measures that can always be used as output measures are product quality (e.g., yield, defects per unit, returns) and product delivery. Lack of performance in these areas affects the customer physically. Price is obviously another measure; however, it does not physically affect the customer the same way as not having a product or having a product that does not work. These measures can usually be established quite easily; the difficult part is determining organizational accountability for the performance. If the company fails to address this alignment, the progress toward continuous improvement will be limited to localized optimization and the operation will miss the big opportunity.

Process-Driven Measures: Infinite Continuous Improvement

There are two other measurements worthy of discussion which are readily understood by the shop floor and can be utilized to drive continuous improvement behavior. One is process cycle time and the other is process quality measured via roll-through yield (Figure 4.2). Roll-through yield is the cumulative performance of each operation in a process. The idea here is that, if a company is building better and better products and delivering them in less and less time, then there should be a positive correlation to total product cost. If you are spending less time reworking defects, replacing scrapped material, moving assemblies around the shop, and waiting for component parts, then overall productivity will improve. When a manufacturing operation allows only first-quality products to proceed to the next operation and does not let material sit around in the shop, then the organization is focusing its efforts on the activities necessary to sustain continuous improvement.

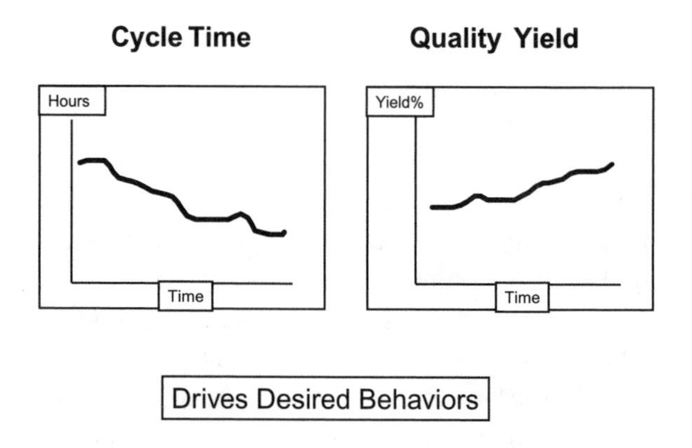

Figure 4.2 Process-Driven Metrics

Goal Alignment Through Policy Deployment

We have met the enemy and he is us! Anyone who has tried to align a bureaucracy and move it in one direction can certainly relate to this statement. Between the business politics, functional silos, misinformation, and lack of information, it can be quite the chore. How, then, does one accomplish this seemingly unachievable task? It can be done through the use of consistent policy deployment. Another term would be *Hoshin planning*, which was originally used by the Japanese (Figure 4.3). The major intent behind policy deployment is to steer an entire organization in the same overall direction. When an entire organization is pulling in the same direction, it is much easier to take corrective action and adjust the course. If a company is pulling in several different directions, not only does it use up a tremendous amount of energy, but it also is more difficult to realign to a new direction.

As a company begins to define its direction through a simple mission statement, it must establish a strategy that achieves that mission. In turn, this strategy becomes supported by specific operational objectives that must be executed and coordinated across the organization. It is at this point that alignment through policy deployment is required. Policy deployment can be compared to requirements cascading down a staircase. At each level within the business, a separate set of objectives and goals can be defined. A statement of the objectives at a division level would be different than those at a department or shop floor level; nevertheless, they can all be aligned to the same company objective. For instance, ABC Company wants to increase market share by 10% in a particular segment of the globe, and they have determined

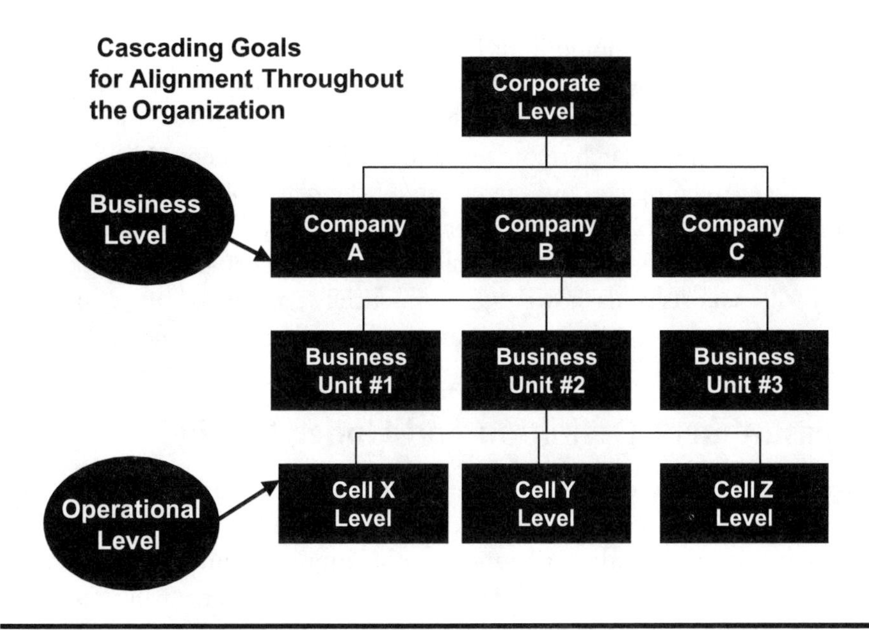

Figure 4.3 Hoshin Planning Process

that the way to accomplish this is through improving the speed of product delivery to the customer. This requires a performance change from a 3-week lead-time down to 1 week. In order to achieve this, manufacturing and purchasing have specific actions they must take that support this company objective. Manufacturing needs to review how product is flowing through the factory today and see where they have waste in the process. They need to verify if the necessary skills and capacity are available to handle the increased volume. Purchasing needs to work with the supplier base to reduce replenishment lead-time so inventory levels do not rise along with the increased volume and to make sure that communication channels for changes in demand are timely.

The demonstrated alignment of these objectives in the planning stages and the constant communication feedback during implementation allow policy deployment to work. It provides total visibility to the organization so that everyone can see where they fit into the success of the company. It focuses the organization on planning the work and working the plan. The regular reporting of progress is necessary in order to understand the current status and to take corrective action. Organizations are a spider web of interrelationships; therefore, it is imperative that each function understands the impact of business changes on the other functions.

An organization should limit itself to three to five company-wide initiatives at any one time. Any more than that leaves a plate that is too full and detracts from the overall focus of the company. There will be too many #1 priorities and not enough resources to cover all commitments. Initiatives will take longer to complete, and the quality of the deliverables will suffer. Keep the plate manageable. As one project finishes up, add a new initiative and drive each to completion. Policy deployment does not guarantee that a company will make its goals; however, it is practically guaranteed that a company will not reach its goals if it cannot even achieve them on paper.

Measurement Definition and Understanding

Individuals are more likely to strive and achieve a performance target they understand as opposed to one they do not. If measures are being posted in a work area by an outside entity and are not completely understood by those being measured, it is not likely that performance in that area will improve. If people cannot describe their measure of performance, do not own that measure, do not report on the measure, nor understand cause and effect relative to the measure, then it is unrealistic to expect the measure to improve. Here, we can draw upon the experience of John G. Belcher, long-time vice president of the American Productivity Center, who stated in his book, *Productivity Plus: How Today's Best Run Companies Are Gaining the Competitive Edge*: "An organization that tries to realize significant productivity improvement without the participation and support of its employees is working against itself. It doesn't make much sense to embark upon a major undertaking when the bulk of the organization misunderstands — or worse yet, resists — the object of that undertaking."[1] In order for a cell team measurement system to work, it is necessary that the metrics be defined, owned, controlled, monitored, and understood by those using the measure (Figure 4.4).

- Created by the Cell
- Owned by the Cell
- Monitored by the Cell
- Controlled by the Cell
- Cause & Effect Understood by the Cell

Figure 4.4 Measurement Objectives

To engage individuals in the improvement process, they must be part of the development of that process. They need to understand where they fit in and how they affect the outcome. It does no good to create a metric in a vacuum, bring it to the floor, provide no definition as to what the measure means, collect the data off-line, have someone outside the area report on the metric, and then expect people to improve their performance.

To overcome this tendency, it is advisable to select a handful of desired outcomes (three to five) and work with the cell team to develop appropriate measures for those desired outcomes. Do not clutter an area with the top 25 measures for that operation. First of all, such postings take up space and get in the way; second, they are not as meaningful to those in the cell. Facilitate agreement among the team about common definitions, identify where the data will come from, select those who will report progress, and establish an expected target performance level. Be sure to provide insight as to how performance of the measure can be improved in relationship to the desired target level. It does no good to expect a target level that no one knows how to achieve.

Be cognizant of the fact that the measurement system that is developed based on the needs of today could change to meet the needs of tomorrow. Measurements will change based on the market, the customer, different levels of performance, and changing competitive priorities. Again referring to *Making Manufacturing Cells Work* by Ingersoll Engineers, change is a constant, and locking into one particular measure today could render a company uncompetitive tomorrow: "Any change in items such as product, delivery, machines, or tooling may well cause changes in the need for certain types of performance measures. Companies move rapidly into and out of markets and otherwise change business strategy to adjust to ever-changing competitive pressures, and existing performance measurements must be continually reviewed in response to these changes."[13]

This chapter has focused on a short list of metrics that can effectively guide an organization on its path to sustained continuous improvement. It has demonstrated how the shop floor can be linked to a company's operational objectives. It has also tried to emphasize the element of measurement that requires an understanding of human behavior and its impact on desired performance. The next chapter begins to reveal where these measures can work as control points in monitoring performance between customers and suppliers.

5 Logistics Element

It is now time to address the element representing the greatest operational challenge — Logistics. This is the area in which all the old rules of operating the shop floor are challenged. This is where the turf wars are fought, functional silos are brought down, individual kingdoms are destroyed, worlds are dominated, universes are lost … well, maybe not quite that big a challenge. Nevertheless, now that responsibility and accountability have been driven down to a lower level within the company, a different set of rules applies and some new techniques will need to be utilized.

This new way of doing business involves changing not only the formal documented process for planning and control, but also the informal, time-tested shopfloor rules that have been ingrained within the organization's culture over the years. Therefore, a lean manufacturing implementation is not only changing documented procedures and physical material handling methods, but it is also placing stress on an informal system that has been used for years. This informal system is usually more difficult to combat.

To appreciate how strong the informal system within an organization can be, ask yourself how quickly and effectively rumors pass through your organization. Enough said. This being the case, it becomes painfully obvious that the communication plan (identified in Chapter 3) is of paramount importance to the success of an implementation. Employees need to understand why their informal system is being challenged and what this impending change will do to affect their work place.

This term *logistics* can mean several different things to different people, so we will clarify its definition here. The term, in this context, refers to those operational elements required to transfer work to a cell, through that cell, and from one cell to the next. It is primarily those in-bound, internal, and out-bound aspects of planning and controlling the flow of work that are

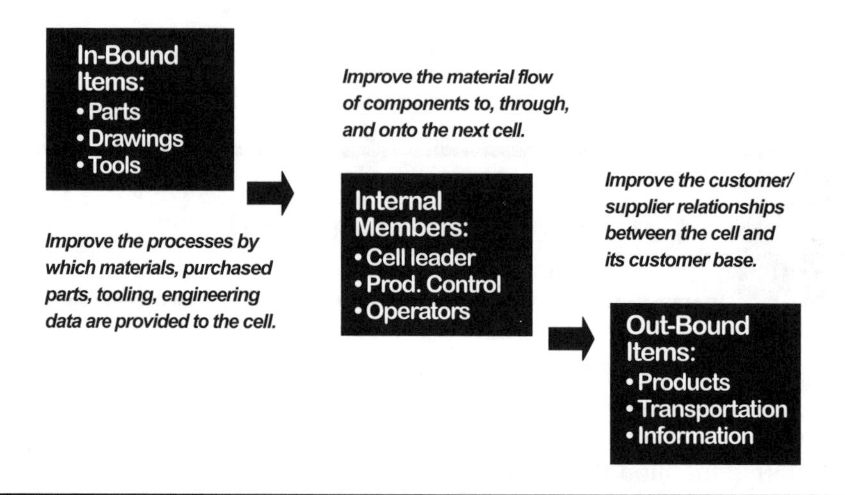

Figure 5.1 Logistics Scope

involved in this element. Following is a description of the scope of each of these aspects — in-bound, internal, out-bound (Figure 5.1):

1. *In-bound* includes all activities related to getting raw material, procured items, and other direct or indirect manufacturing items to their respective places of consumption. Functions such as procurement and subcontract management and items such as engineering drawings, process specifications, and tooling are all associated with in-bound logistics.
2. *Internal* has to do with those items required to facilitate the flow of work through the cell. These items involve cell team members (e.g., cell leader, production engineer, shop touch labor, production control) and include such physical elements as materials or purchased parts, production tooling, equipment, Kanbans, priority listings, etc.
3. *Out-bound* relates to those items required to exit from the supplier cell and arrive at a customer or customer cell. Items such as customer identification, a negotiated delivery quantity, kit definition, supplier-held inventory, mode of transportation, ownership exchange points, etc. are all areas of focus for this aspect.

Now that a general idea of scope and boundary has been established, the various principles involved with the logistics process can be explored individually. Our focus in this chapter will be on:

1. Planning/control function
   a. Priority planning (forward plan)
   b. Capacity planning (workload)
   c. Capacity control (input/output control)
   d. Priority control (dispatch list)
2. A,B,C material handling
3. Service cells
4. Customer/supplier alignment
5. Just-in-time (JIT) Kanban demand signals
6. Cell team work plan
7. Level loading
8. Mix-model manufacturing
9. Workable work

Planning/Control Function

The planning/control function exercised within a cell can go by several names (e.g., constraint scheduling, release and control, workflow management) and yet still mean the same in terms of functionality. The planning/control function described here requires that specific work rules be utilized during the operation of a cell. Remember that the Logistics element has as much to do with changes in work rules as it does with physical process changes. Examples of some of the standard operational work rules may include:

- Do not load the cell equipment over 90% of demonstrated capacity.
- Do not release work inside average actual lead-time.
- Release only workable work to the cell.
- Sequence work by using first-in/first-out prioritization.
- Do not release work without an authorizing Kanban.

These are operational work rules that are to be developed, defined, and documented by the cell team members. Through the education and training (received via the Organization element), the cell team will have a better understanding of the need for new work rules, and, because they have developed and defined those rules, there is greater ownership for them as a team. These work rules are not intended so much to reduce flexibility as they are to facilitate consistency, structure, and continuity among team members for operations of the cell. This collaborative approach to work rules in actuality will enhance both the responsiveness and predictability of cell performance, especially within a multi-shift environment.

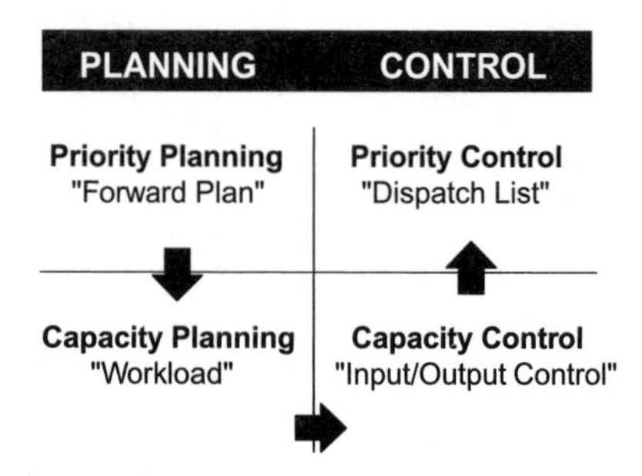

Figure 5.2 Aspects of Planning and Control

Planning and control are critical functions that contribute to the successful implementation of the cell. Often times, those individuals only focusing on utilizing just-in-time (JIT) and Kanban material pull overlook these functions. The initial implementation and subsequent day-to-day operations of a cell are greatly influenced by: (1) how executable the plan is, and (2) how robust the control mechanism is. There exists a strong relationship between these functions, as the better the planning effort, the easier the control effort. The criticality of this relationship was emphasized by the Japanese master of production engineering, Shingo Shigeo (*Non-Stock Production: The Shingo System for Continuous Improvement*): "If the planning level is about 80 percent, control precision need only be around 20 percent. If the planning level is about 50 percent, control precision needs to be around 50 percent."[20] There are four basic aspects to planning/control within a cell (Figure 5.2), and each is explained in detail in the following discussion.

Priority Planning (Forward Plan)

This aspect is concerned with planned or future workload requirements, which are normally fed to the cell by a manufacturing resource planning (MRP II) or some other requirements planning system. Lean manufacturing in no way abolishes the need for requirements planning; rather, it actually requires it in order to: (1) establish cell design criteria, (2) plan short-term workloads (1 to 4 weeks), (3) perform make/buy analyses, and (4) communicate future demand needs to upstream suppliers. A company's current requirements planning system is usually adequate enough to provide the required information for a lean manufacturing environment.

Capacity Planning (Workload)

This function is necessary for the cell team to review and agree upon the upcoming workload, manpower, and overtime requirements necessary to satisfy customer performance expectations. This capability allows the cell team to determine its own destiny and provide input into the decision process that controls end product performance. The cell level visibility to future workload fluctuations can then be mitigated by the cell team through level loading, off-loading, lot size splitting, planned overtime, etc.

Capacity Control (Input/Output Control)

The cell team is held accountable to manage performance to plan. Capacity control is used as a control device to provide the cell team with the capability to maintain workload visibility and monitor progress to plan. They are given the opportunity to take credit for achieving an operational goal or are provided with the ability to take swift corrective action when performance is falling off the mark. Managing queue sizes is paramount to meeting lead-time commitments. If actual queues are exceeding plan, then promise dates to customers will be missed and customer confidence will diminish. Lead-time variability (a true menace to many delivery problems) is a direct reflection of how well actual queue times are kept in control.

Priority Control (Dispatch List)

The sequence by which work is introduced to the cell will be a function of three things: (1) Is there a customer demand? Even though there is a planned requirement for an item, until there is a demand pull signal from the customer, there is no real need for the item. (2) Is there enough capacity? Until capacity has been cleared or a Kanban container becomes available to introduce more work into the cell, it cannot release work. If work was released, work in process would increase beyond the cell design parameters, queues would grow, and lead-time would increase. (3) Is the work package available? Unless all the items necessary to work a job have been made available, work cannot be released to the cell. If incomplete work packages were released to the cell, they would eventually stop. They then would have to wait for resources to be applied to break it loose, and the workflow would begin to backlog.

A,B,C Material Handling

In a lean environment, controlling the flow of material and managing inventory will change under the new operating rules for material handling. Instead of managing each and every part exactly the same way, parts will be reclassified based on their demand behavior characteristics. For instance, a large, complex machined part weighing 500 pounds with 100 hours of machine time would be scheduled and controlled differently than a nickel/dime clip or bracket. The amount of time, money, and resources required to manage inventory should be comparable to the behavior characteristics of the part or components. Parts should be stratified according to a given criteria so that an appropriate amount of effort is expended on managing the part replenishment process.

Parts or components can be segregated along an A,B,C type of classification (Figure 5.3). This approach differs slightly from Pareto's 80/20 rule; however, the average part population still falls along the normal 15/35/50 percentage split. For example, parts that are expensive, more complex to build, and often exhibit long lead-times should be considered "A" parts. They should be scheduled with suppliers either (1) with transportation pipeline Kanbans (especially with high-volume product), or (2) directly through MRP II (for low-volume product), just as in most plants today. "B" parts are usually less complex, have shorter and more predictable lead-times, are less expensive, and are small enough to be kitted (if required). These can be replenished via Kanbans and can possibly be built on demand. These parts could be built

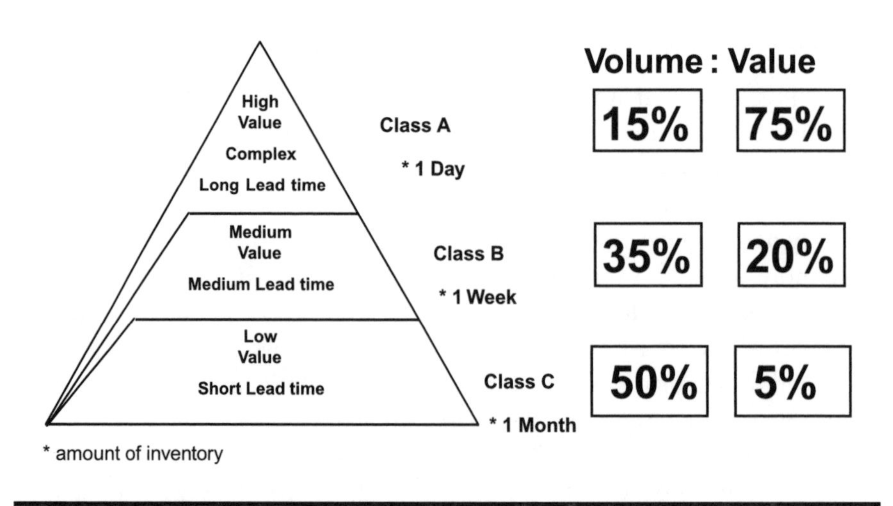

Figure 5.3 A,B,C Material Handling

and delivered in negotiated batch sizes or in predetermined kits (if required). If the demand volume is low or highly variable, it may make more sense to replenish these parts via MRP II or through nonrepetitive Kanbans. By far the majority of parts (50%) would find themselves in the "C" category and could be managed directly through a vendor-managed reorder point or Kanban system. These parts would appear on the bill of material, but would not be scheduled per MRP II and therefore would require minimal manpower to manage.

Service Cells

In an ideal world, all parts would be manufactured complete within a given cell. All the necessary manufacturing processes would be located in that cell and the parts would never have to leave the cell. Raw material would come in and a completely finished part (ready for consumption by the customer) would go out the other side. Now, if you currently have this scenario operating within your plant, then pass by this section because it does not apply to you. However, if you are like the majority of the manufacturing community, you certainly do not have enough capital to fully populate your cells in this manner. This section will provide an option for your facility.

Have you ever been to a drycleaner with a load of shirts and read the sign out front, "In by 9, out by 5?" Have you ever been to a train station and ridden on a train? Did you notice how the conductor continually checks his watch and monitors the time in the station? He is making sure the train enters and leaves the station on time. At 8:00 a.m., for example, announcing "all aboard" indicates that the train is leaving the station. Anyone there can board, and those who are not there will have to wait until the next scheduled train arrives. What if certain capital-intensive manufacturing operations were set up to run in the same manner? The above-mentioned scenarios describe two types of rules that can apply with service cells (Figure 5.4). These service areas are designed to support cells that are manufacturing products. Their objective is to satisfy the needs of the manufacturing cells and to provide a predetermined level of service or turnaround for a particular process. Because, as we learned earlier, manufacturing cells are accountable for the product from cradle to grave, they become highly dependent on service cells to provide consistent, predictable process turnaround. This level of dependency strengthens the customer/supplier relationship and ties in directly with the Metrics element that we explored in Chapter 4.

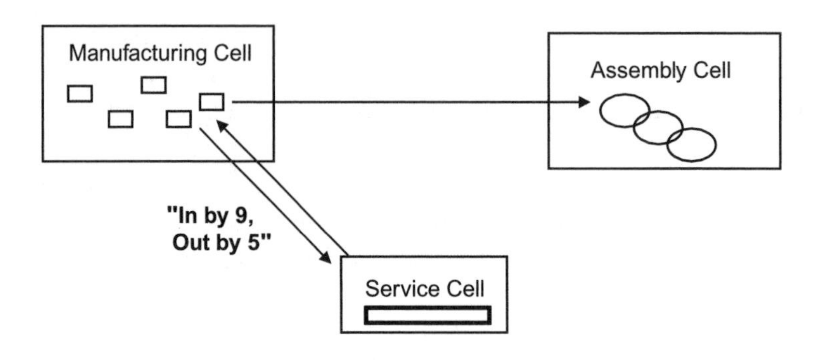

Figure 5.4 Service Cell Relationship

Customer/Supplier Alignment

When companies pay homage to the terms *customers* and *suppliers*, it is normally the type of "motherhood and apple pie" lip service that does not mean much in the way of substance. In order for lean manufacturing to truly function, direct lines of communication between customers and suppliers must be identified and strengthened. For every product produced within a manufacturing cell, there is a corresponding customer or customer cell that will be consuming that product. Whether the supplier cell is part of an internal customer/supplier relationship within a multi-plant facility or part of a larger supply chain involving several different companies, the same adage applies: alignment with the customer. Cell members should recognize who utilizes their parts and know if those parts are satisfying the customer's fit, form, and function requirements. Are they packaged correctly? Is there a better kitting procedure that could be utilized? Can we negotiate a better delivery quantity to help our total product cost? Who do they call when there is a quality problem with the last parts that were received? These are all legitimate questions that can be asked and answered when there is direct alignment between customers and suppliers.

One way to begin establishing this alignment relationship is to:

1. Run a "where used/received from" list off the bill of material for all parts/components that are assigned to a given cell.
2. Sort the parts by four categories: customer, volume, cost, and destination, which allows prioritizing investigative efforts.
3. Call on the biggest customers or suppliers first to assess their needs and begin negotiating ways of improving the supply chain.

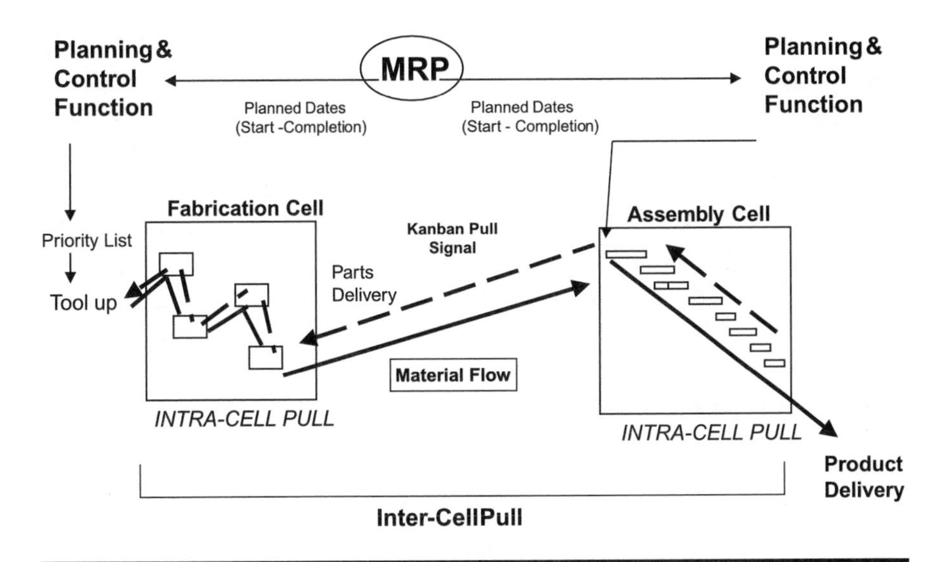

Figure 5.5 Just-in-Time (JIT) Demand Pull Signals

In addition to capturing the part numbers, a process map of the activities for the cell can be very useful. By utilizing a format of supplier-input-process-output-customer (SIPOC), a great deal of information can be obtained in regard to input requirements for the process and output requirements of the customer. By capturing the key activities within a process (e.g., a cell, supplier interface, shipping, order administration) and categorizing them according to value-adding or non-value-adding, significant insight into the performance of a process can be obtained. Remember that this is documenting activities, not tasks. Keeping the process map at the appropriate level of detail can be difficult. Activities are focused on the verb-noun (action to an object) relationship of functions in a process, while tasks are more the procedural-level steps for those activities. To keep this distinction straight, try using the guidelines set by Colkins in his *Activity-Based Cost Management: Making It Work: A Manager's Guide to Implementing and Sustaining an Effective ABC System*: "A good rule is to think of activities as what people do and the tasks that make up activities as how the people perform activities."[2]

Just-in-Time (JIT) Kanban Demand Signals

A multitude of methods can be exercised when utilizing pull signals (Figure 5.5). They include cards, standard containers, empty squares on the shop floor, electronic messages (e.g., faxes, e-mails, EDI, e-commerce), among others. All of these methods have several aspects in common. First, the demand signal represents the authorization to begin work. Second, no job is to be released for work without a demand authorization from the customer. Third, the demand signal controls the amount of work in process allowed in the supply chain. Fourth, the number of Kanbans in the system will determine the amount of work in process for the chain. This scenario gives the cell the ability to control variability in lead-times, as queues are unable to grow beyond the number of calculated Kanbans. Fifth, no one is allowed to knowingly pass defects on to the next operation; defects are to be stopped when they are found and corrective action incorporated immediately. Sixth, workflow is prioritized on a first-in/first-out (FIFO) basis. This not only applies to the scheduling of work into the cell, but also the physical handling of material. The physical inventory turnover of material is just as important as the financial inventory turnover of material. These aspects of managing Kanbans are summarized in the following rules:

1. A Kanban demand signal is the authorization to begin work.
2. No job is to be released without demand from the customer.
3. The Kanban controls the amount of work in process allowed in the flow.
4. The number of Kanbans will control the manufacturing lead-time through queue management.
5. Do not pass known defects on.
6. Utilize first-in/first-out (FIFO) material flow.

Kanbans can be set up between workstations, between workstations and point-of-use (POU) locations, between cells and central stores, between assembly cells and fabrication cells, between fabrication cells and external suppliers, and between assembly cells and customers. Each relationship will have its own individual issues to address as to location, size, quantity, ownership, shelf life, weight, etc. For the purposes of general discussion in this section on methodology, there will be two types of Kanban material pulls, one depicted as intra-cell (internal to the cell) and one as inter-cell (external to the cell). Each has a relationship with production cells and the overall objective for Kanban demand signals.

The type of Kanban methodology deployed is very dependent on the manufacturing environment, the receptivity to change of an organization's culture, and a supplier/customer's motivation for participation. The more difficult the environment, the more robust a process required. No one knows the manufacturing environment better than each individual company; therefore, each company is in the best position to determine which method to use. However, no matter what method is chosen, the six rules for managing Kanbans still apply.

Cell Team Work Plans

As was mentioned in Chapter 3, a lean manufacturing cell operates as a team completely focused on delivering a product to a customer. For this team to function as one cohesive unit, they must agree on how they will operate. Earlier, in the planning/control section, the idea of a forward plan was introduced. This plan provides a future look at the production requirements coming to the cell over the week. The cell team reviews this forward plan, ensures that they have enough capacity and resources to make this plan (if not, they will make the proper adjustments), and agree, as a group, to execute this plan. This way there is buy-in to the schedule by all the team members. They own the performance objectives for the next week and they have developed synergy around the plan.

This review process should take place on a regular basis (e.g., weekly) and become part of the routine management of the cell. The cell leader should anticipate facilitating this discussion, and the support personnel should plan on performing an analysis on the data before presenting it at the meeting. The meeting can then move along efficiently and with little wasted effort. This may appear to be a simple, common-sense activity, but it is surprising how many cell implementations never utilize this activity and later wonder why the cell teams are not achieving the targeted objectives and are floundering without a common focus.

Level Loading

According to one of the leading authorities on supply chain management, William C. Copacino, in his book *Supply Chain Management: The Basics and Beyond*, there are four prerequisites or pillars required for a JIT system to function properly: "If JIT logistics plans are to work, four pillars must be in place … stable production schedules, efficient communication, coordinated transportation, quality control."[4] It is one of these prerequisites — a relatively level production schedule over a defined period of time — that is the subject of this section. In order to align customer demand with takt time (see Chapter 6), a need exists to level demand at a rate that is conducive for both the supplier and the customer.

	wk1	wk2	wk3	wk4	wk5	wk6	wk7	wk8
Booked Orders	150	155	170	150	140	165	160	155
Forecast	125	140	165	135	155	125	130	130
Rate Production	150	155	170	150	155	165	160	155

Daily rate schedule month 1
150 + 155 + 170 + 150 = 625 units
625/21 days/month = 30 units/day

Daily rate schedule month 2
155 + 165 + 160 + 155 = 635 units
635/20 days/month = 32 units/day

Figure 5.6 Rate-Based Schedule

By presenting the customer-forecast information in units per day or week, an understanding as to the demand pattern and volume variation for a given set of products can be analyzed. This information provides insight into the development of a level rate-based schedule for a production cell. This level rate-based schedule of demand over a given period of time is only for products that fit a rate-based demand pattern which demonstrates a relatively high volume of demand, a consistent customer order frequency, and limited volume fluctuation. By utilizing a rate-based schedule (Figure 5.6), these products are scheduled less often (e.g., once a month) and are designed to be produced at a given rate for a given period of time. Realizing that demand does change, customer demand patterns should be monitored on a regular basis and the scheduled rate adjusted accordingly.

To develop a rate-based schedule, take the forecast information in units by day or week and compute a monthly average. Then, compare the forecast monthly average to the booked orders and develop a rate of production from the higher of the two numbers. This is done in order to buffer against variation in customer demand. This methodology is similar to the total demand process talked about by Costanza in *The Quantum Leap: In Speed to Market:* "The definition of total demand inside the Demand time fence is the sum of actual customer and finished goods replenishment orders. Total demand outside the Demand time fence is the greater of the forecast and actual customer orders."[5]

From this point, the production cells can commit to a production schedule on a weekly basis and be held accountable for achieving their planned output. As was stated earlier, not every product has a demand pattern conducive to rate-based scheduling, but for those that do, this can be a very effective methodology.

Mix-Model Manufacturing

There are manufacturing cells with workstations that can be designed to produce a variety of products and volumes over a given time frame. These cells are capable of performing what is known as mix-model manufacturing (Figure 5.7). The criteria for designing these types of cells requires that the production processes be relatively consistent from part to part without a significant amount of variation in the process. In addition, these cells usually contain a highly flexible workforce, have limited variation between work content times for each operation, and can change over between products very rapidly.

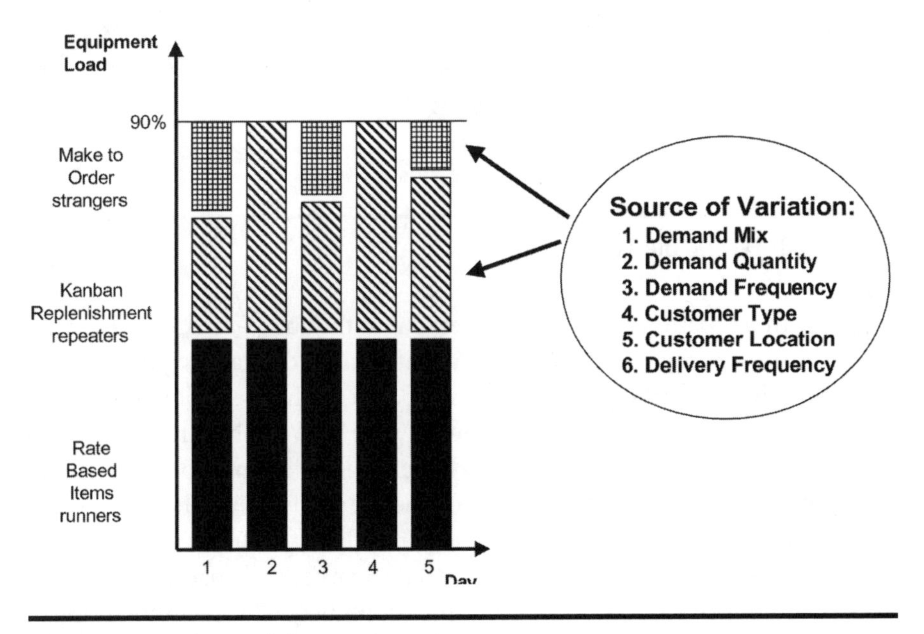

Figure 5.7 Mix-Model Manufacturing

Mix-model manufacturing provides the ultimate responsiveness and utilization of floor space. It supports making any mix of any product on any day (provided the products were designed for the cell). Again, Schonberger, in his book *Japanese Manufacturing Techniques: Nine Lessons in Simplicity*, described the positive effects of mix-model manufacturing: "An advantage of mix-model sequencing is that each day you make close to the same mix of products that you sell that day. This avoids the usual cycle of a large buildup of inventory of a given model, followed by the depletion to the point of potential lost sales as the next model builds up."[15]

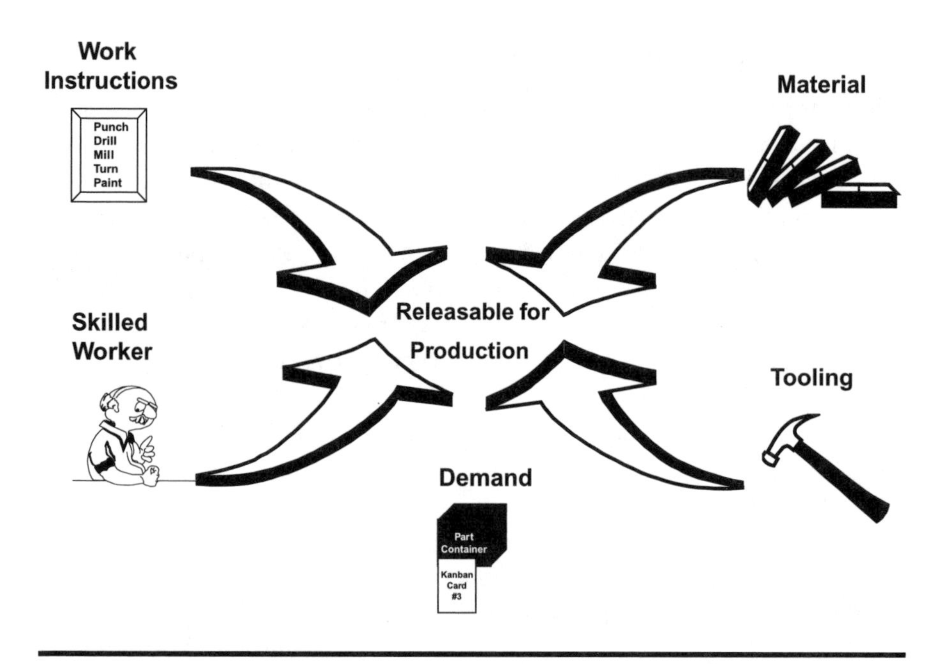

Figure 5.8 Workable Work

Once those manufacturing processes that fit the above-mentioned criteria are grouped together in a cell, it is just a matter of understanding the product demand behavior and segregating the mix based on that behavior. Rate-based products are made in the same amount every day. Kanban replenishment products (often finished-goods stock) are replenishments for Kanbans as demand requires. The make-to-order or special products will be made when there is enough capacity remaining to produce those products. By scheduling product this way, one makes the most effective use of space, equipment, people, time, material, etc. The concepts of runner, repeater, and stranger, which are applicable to this methodology, will be discussed in greater detail in Chapter 6.

Workable Work

Workable work is a term that refers to those elements contained within the manufacturing process that are necessary in order for work to begin on a product (Figure 5.8). Every manufacturing environment will have something that is specifically required in order to begin work; however, all environments will have the following elements in common: (1) material, (2) tooling, (3) work instructions, (4) demand, and (5) skilled workers. Most MRP II systems are set up to plan and release work to the shop floor based on demand information generated from the system. Some have a logic setup to check for component part availability before assembly orders are launched, but that is normally where it stops.

The problem that arises in many plants is that work is released to the shop floor without having verified completely that it is workable. For example, work order IS1234 is launched to the first operation. The part is blanked and moves on through operations two through five; however, when it arrives at operation six, there is an issue. The tooling is out for repair and not available to run this job. What happens? The job sits and waits until the tooling is available. This happens every day in plants, and the more complex the manufacturing operation the more this launch-and-wait behavior is evident. Instilling an awareness of the concept of workable work sets in motion a process that verifies the availability of those critical elements required by manufacturing, before committing a job to the shop floor, thereby eliminating the delays and wait time that are so indicative of long manufacturing lead-times.

The topic of logistics is a very broad subject matter that could easily fill an entire textbook. This chapter was only intended to cover some of the primary aspects that should be addressed as part of a lean manufacturing implementation. Now that we have a greater appreciation for the infrastructure elements, it is time to address the element that is most familiar to people — Manufacturing Flow.

6 Manufacturing Flow Element

Most practitioners within the field of manufacturing can relate to tangible, hard-fact types of projects that individuals can go and lay their hands upon, so to speak. These are the type of improvement initiatives most readily embraced and implemented. These projects are the most visible, and they are witnessed by everyone within the organization. This is why the idea of rearranging equipment and altering shopfloor layouts is pursued so passionately by many manufacturing organizations. Improvements are easily recognizable, and it is obvious that change has taken place. In order to win this particular crowd's acceptance for a holistic approach to lean manufacturing, hard-fact results must be evident. This being the case, this chapter presents a series of cell design techniques based on hard-fact material which should be utilized when deploying a lean manufacturing concept similar to the one described in this book.

The following series of techniques is to be used when assessing products and their associated process flow and translating that data into usable information for generating a cell design:

1. Product/quantity (P/Q) analysis (product grouping)
2. Process mapping
3. Routing analysis (process, work content, volume matrices)
4. Takt time
5. Workload balancing and one-piece flow
6. Cell design guidelines
7. Cell layout
8. Kanban sizing

Product/Quantity Analysis (Product Grouping)

The first step in this process is to gather and understand product demand data (Figure 6.1). This is accomplished by generating a cumulative Pareto percentage, by volume, of all product stockkeeping units (SKUs). These data items originate at the customer and provide a baseline by which to begin demand behavior analysis. Annualized product SKU demand data should be segregated on a monthly/weekly/daily demand basis. The source of this information usually comes from the business plan forecast (in units) and covers a time horizon of 6 to 12 months. By displaying the cumulative percentage, both high- and low-volume products begin to present themselves. In addition to the forecast data, it is important to consider the actual customer order sales data. Doing so accounts for actual demand volume and mix variation, which is important input for the takt time calculation (discussed later in this chapter).

Product Group	Product SKU	Qty/Yr	% Cum	Qty/Mth	Qty/Day	Run Cum
1	H34967	7500	25%	625	31	25%
	F34786	6727	21%	560	28	46%
	H34987	5540	17%	460	23	63%
2	U89756	2302	7%	192	10	70%
	H56097	2300	7%	192	10	77%
3	H89255	1829	6%	152	8	83%
	P94565	1300	4%	108	5	87%
	E34578	899	3%	75	4	90%
4	H87566	230	2%	19	1	92%

Figure 6.1 Product Demand

The P/Q analysis (Figure 6.2) looks for natural breaks in product groupings by sorting the gathered data and determining a fit for production cells by: (1) their associated volumes, and (2) product alignment characteristics. This is usually an iterative process and is conducted several times in order to determine a best fit for each cell type. Product alignment characteristics could include the following criteria:

1. Align high-volume products together.
2. Align to specific customers, such as original equipment manufacturers (OEMs).
3. Align to specific target markets.
4. Align to common manufacturing processes.
5. Align to configuration commonality (size, material, function, etc.).
6. Align to engineering content (standard vs. special).

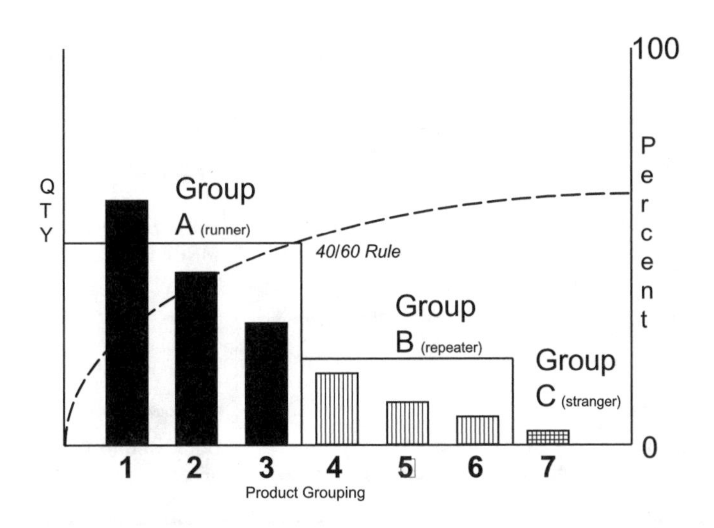

Figure 6.2 P/Q Analysis

After the products have been identified and segregated into product groupings, they are sorted by volume and plotted on a chart. This chart visually displays the natural breaks in volume by group. Normally, 40% of the products will account for 60% of the part volume (this is referred to as the 40/60 rule). When a product grouping falls into this category, it is wise to establish dedicated flow lines/cells with segregated resources in support of this product grouping. These products are called *runner* products because they have high volumes, frequent customer orders, and stable demand (Figure 6.3). The remaining balance of product groupings will fall into one of two categories. The first group fits a general purpose or flexible cellular operation known as *repeaters.* This category has a greater variety of products, which will be produced across resources that are not dedicated to a specific flow line. Parts that have lower volume amounts, variable order frequency, and/or high variability in operational routings will find their way into this category. The second category is that of *strangers.* This category is for miscellaneous items that are being produced within the plant as one-off items or that have a very low volume or infrequent (once per year) demand pattern. These items are usually best managed through MRP II and can be segregated from the rest of the factory by:

1. Establishing separate production area
2. Running the products once or twice per month
3. Running them when capacity is available
4. Outsourcing the products

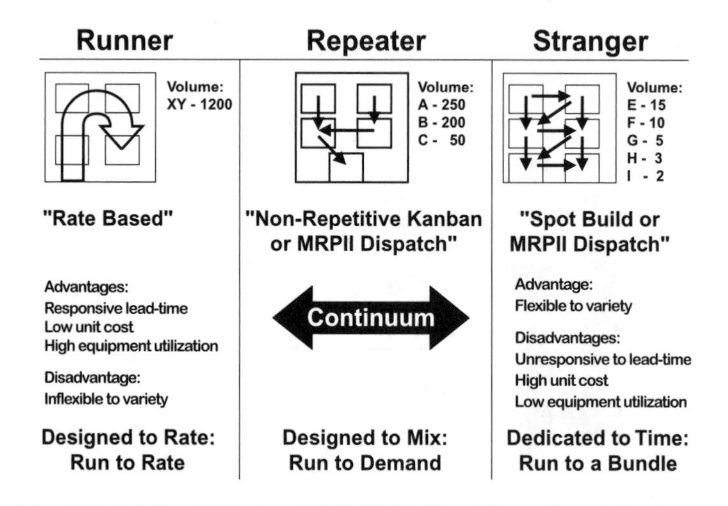

Figure 6.3 Runner, Repeater, Stranger

5. Running them once per year and holding in finished-goods stock
6. Making one final run and eliminating the item from current product offering

Hill, in his *The Essence of Operations Management*, addresses this same idea of segregating products, only his focus is on the market place: "Companies need to recognize that low-, medium-, and high-volume batch processes handle a very wide range of volumes with correspondingly different order-winners. For companies, therefore, to assume that the choice of one process, even for a single category such as batch, will provide support for the level of diversity associated with a normal range or products/services is a mistake."[10] In either case, whether by product alignment criteria or order winners, it is important to recognize that all products are not demanded the same and therefore should not be managed the same.

Process Mapping

Once the product demand behavior is understood, the next area of analysis is that of process mapping. It is necessary to know what operations are required to produce the products being considered for cell design. In the end, the final design of the cell will need to account and accommodate for all process steps, whether accomplished in the cell or not. Block process mapping (Figure 6.4) is usually conducted on the highest volume products. The lower

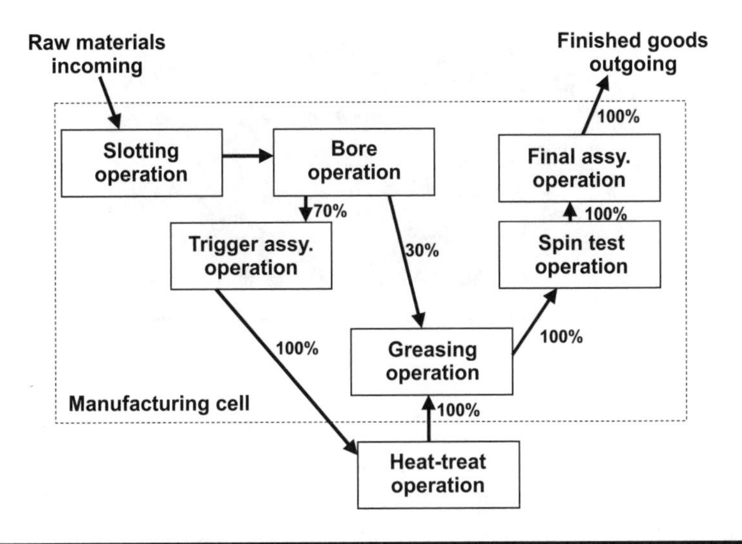

Figure 6.4 Block Process Mapping

volume products can be mapped separately if the process steps are significantly different; however, this is usually not the case. By actually walking the process, documenting the steps, and talking with the process owners, a good representation of the product flow and volume can be documented.

In addition, to the block process map, a spaghetti diagram (Figure 6.5) is created in order to grasp the magnitude of operator and material travel in the current process. The reason it is called a spaghetti diagram is that by the

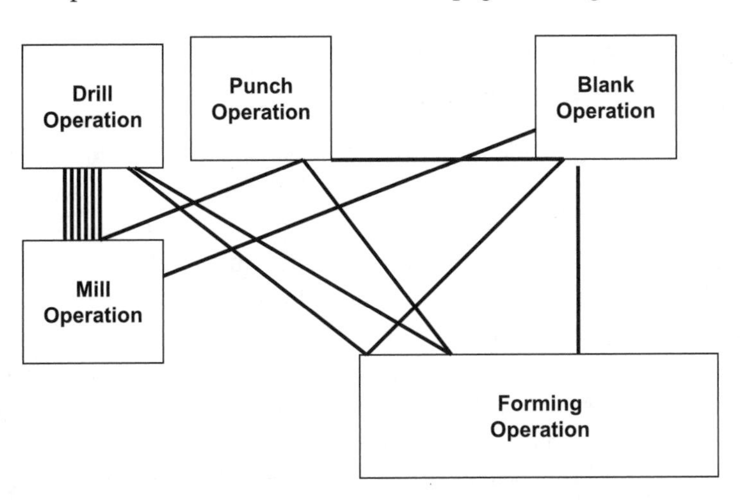

Figure 6.5 Spaghetti Diagram

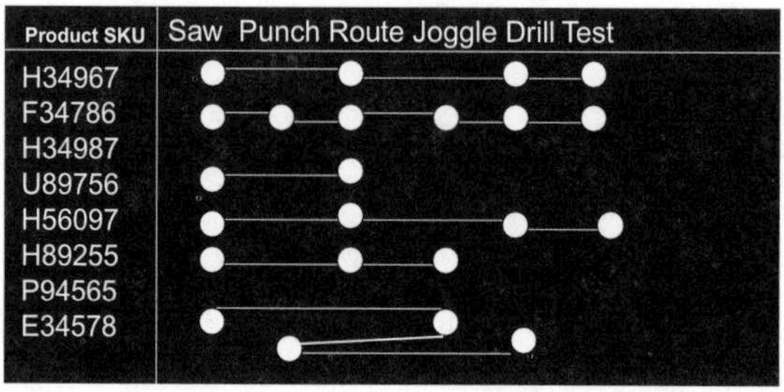

Product SKU | Saw Punch Route Joggle Drill Test
H34967
F34786
H34987
U89756
H56097
H89255
P94565
E34578

Figure 6.6 Process Matrix

end of the exercise of recording the current process the drawing looks like a bowl full of spaghetti. This simple technique is nothing more than drawing the area under assessment, representing the operator and material movement on paper with a colored marker, and measuring the amount of feet traveled for both. What we can learn from this technique is very illuminating. It is not uncommon for an operator to be traveling up to half a mile every time there is a need to gather tools and parts to conduct a machine changeover.

These two tools are used as sources of input during the cell design process. They do a good job of capturing the "as is" condition and visually displaying what is actually happening in the process today. They identify significant opportunities for waste elimination or reduction and provide real data by which to make decisions, rather than relying upon "I think…" or "I feel… ."

Routing Analysis (Process, Work Content, Volume Matrices)

Routing analysis is nothing more than the assessment of workflow patterns and volume/process variation. The first step in this analysis is the creation of a process matrix (Figure 6.6). This is accomplished by placing the routings for each part of a product on a grid. By identifying all manufacturing processes across the top and listing products down the side, a grid is created where each part routing can be physically drawn. By displaying the workflow in this manner, it is easy to see patterns of commonality, resource consumption, and reverse part flow activity. Each of these items is an important factor to consider when establishing cell configuration.

The purpose of a work content matrix (Figure 6.7) is to gather relevant man time, machine time, and setup time for a particular part population. After being collected, this baseline information should be loaded into a database as

MN = Man Time
MC = Machine Time
S/U = Setup Time

Product SKU	Table Saw			Gang Punch			Pin Route			Joggle			Total		
	MN	MC	S/U	MN	MC	S/U	MN	MC	S/U	MN	MC	S/U	MN	MC	S/U
H34967	1.0	1.0	25.0				1.5	1.5	15.0				2.5	2.5	40.0
F34786	2.5	2.5	20.0	3.4	3.4	15.0	0.5	0.5	35.0	4.6	4.6	10.0	11.0	11.0	80.0
H34987	1.0	1.0	20.0				0.5	1.5	20.0				1.5	2.5	40.0
U89756	0.5	2.5	25.0				0.7	1.5	15.0				1.2	4.0	40.0
Total	5.0	7.0	90.0	3.4	3.4	15.0	3.2	5.0	85.0	4.6	4.6	10.0	16.2	20.0	200.0

Figure 6.7 Work Content Matrix

reference data for utilization during the cell design process. This database of information can be gathered in one of at least two different ways. The first is strictly a manual process in which an industrial engineer will conduct a work element analysis and complete a time observation form (TOF; Figure 6.8). The

Process Observed	Observer			Date of Observation							
Operation Steps	1	2	3	4	5	6	7	8	9	10	To tal Avg
1.0 Table Saw	0	0	0	0	0						0
	10	11	11	10	10						10
2.0 Gang Punch	10	11	11	10	10						10
	20	22	22	20	20						22
3.0 Pin Route	8	10	8	10	8						9
	28	32	30	30	28						29
4.0 Joggle	8	8	8	8	10						8
	34	40	38	38	38						37

Average Cycle Time

Figure 6.8 Time Observation Form

engineer will walk the process map for each part and record the actual operational data for each of the process steps. The engineer will need to keep track of both the individual operation time and the cumulative running total time. Depending on the operation, a series of five to ten recordings should be enough to accurately reflect the process. In addition to the time study, interviews with the process owners can provide valuable insight into the existing process flow.

At times, with certain work environments, these "time studies" can be viewed in a negative light, and participation by the shop floor can be difficult to obtain. It should be explained that these observations are being used to try to understand the current process and are not being used to set pay rate. If there is still opposition, then try to reach a consensus utilizing the existing work element standards. In many cases, there is so much improvement opportunity available without being concerned with changes to the actual work content of the process that this is not an issue.

A second approach would be to capture data from the existing MRP II system. This approach is probably more acceptable when trying to address a large population of parts in a short period of time and information accuracy of 95% is not required. If, however, the root cause of a problem is critical or an accurate story is required, then it is recommended that the analysis be performed on data collected directly from the shop floor. This way the engineer can not only formally record what is seen but also record informally what is heard through conversations with shopfloor personnel.

At this point, the work content of the products has been recorded and some insight into the product process flow has been documented. Now it is time to understand the relationship between the two. By reflecting the product and process flow in a volume matrix (Figure 6.9), decisions about the cell design begin to evolve. The volume matrix reflects demand and process flow data in production units and minutes/hours. The placement of product groups on a matrix allows for the calculation of total volume by units and hours for each product and each process. Depending on the manufacturing process, the production rate could be calculated in days or weeks. The hours should reflect three major categories: man time, machine time, and setup time (the setup time being assumed at once per day). One of the objectives of a lean manufacturer is to be flexible and responsive; therefore, the goal is to make today what is sold today. This cannot be accomplished if changeovers are executed once per month.

Again referring to Costanza's *The Quantum Leap: In Speed to Market,* the author describes a similar process of searching for commonality: "Each product is reviewed to identify the particular processes or machine operations required to manufacture each product. The next step in cell design is to create

Product	Table Saw			Gang Punch			Pin Route			Joggle			Total		
SKU (Vol.)	MN	MC	S/U	MN	MC	S/U	MN	MC	S/U	MN	MC	S/U	MN	MC	S/U
H34967 (31)	31.0	31.0	25.0				46.5	46.5	15.0				77.5	77.5	40.0
F34786 (28)	70.0	70.0	20.0	95.2	95.2	15.0	14.0	14.0	35.0	128	128	10	308.0	308.0	80.0
H34987 (23)	23.0	23.0	20.0				11.5	34.5	20.0				34.5	57.5	40.0
U89756 (10)	5.0	25.0	25.0				7.0	15.0	15.0				12.0	40.0	40.0
Total (Min.)	129	149	90	95.2	95.2	15.0	79	110	85	128	128	10	432	483	200

Figure 6.9 Volume Matrix

a cell configuration that is made up of the common machines or operations identified in the process map."[5]

There are two primary outcomes of the routing analysis exercise: (1) the segregation of high- and low-volume products based on a reflective view of the manufacturing process, and (2) an understanding of the degree of variation in product volume/mix and work content as it relates to cell design. It is through an understanding of these two aspects that cell design decisions can be made relative to the use of:

1. Scheduling methodology — complex mix vs. segregated production
2. Rate-based, Kanban, make-to-order products
3. Kanban buffers for line imbalances and long setup times
4. Equipment workloads
5. Equipment needs
6. Staffing needs
7. Shift hour requirements

Takt Time

The word *takt* comes from the German word for rhythm or beat. Takt time is the basis for cell design and represents the rate of consumption by the marketplace (Figure 6.10). Takt time is where the effort starts, because it is reflective of the customer demand. Everything in cell design is based on takt time. Takt time is often confused with cycle time. The two are calculated from completely different perspectives. Cycle time represents the current capacity/capability of the existing operation, whereas takt time is based on projected customer demand, not the ability of the current process to perform. The ratio for takt time has scheduled production time available as the numerator and designed

$$\text{Takt time (TT)} = \frac{\text{Total time available per day}}{\text{Designed daily production rate}}$$

Figure 6.10 Definition of Takt

daily production rate as the denominator. For instance, scheduled time available would be nothing more than a regular 8-hour shift minus time for scheduled lunches, breaks, meetings, etc. This results in the scheduled time available. For example, an 8-hour shift – (30 minutes for lunch + 30 minutes for two 15-minute breaks) = 7 hours of shift time available.

The factors that go into developing the designed daily production rate include the business plan sales forecast and a variation coefficient to cover customer demand mix/volume variation. The combination of these factors result in a designed daily production rate for the cell. For example, a forecast demand might be

	Monday	Tuesday	Wednesday	Thursday	Friday
Units:	200	280	265	215	245

In order to accommodate the volume variation and design a level production schedule, the cell-designed daily production rate would be at 290 units per day. This would be based on reviewing the demand variation from day to day or week to week, determining the average demand for the next 6 to 12 months, and increasing the demand level to accommodate fluctuation by a coefficient. In this case, the average demand is 241 units plus a 20% coefficient, or a daily demand of 290 units (see below). The percentage is subjective, based on the amount of variation; however, it is not recommended to exceed 50% of the average because a cell cannot be designed for infinite capacity.

$$200 \text{ units} + 280 \text{ units} + 265 \text{ units} + 215 \text{ units} + 245 \text{ units} = 1205 \text{ units}$$

$$1205 \div 5 = \text{average of 241 units per day}$$

$$\text{Variation coefficient} = (280 - 241) \div 241 = 17\% \text{ (rounded to 20\%)}$$

$$(241 \times 1.20) = 290 \text{ units designed daily production rate}$$

The takt time for the example above would be 1.5 minutes. The time available in minutes is 7 hours × 60 minutes, or 420 minutes. Dividing 420 minutes by 290 units gives 1.5 minutes, which is the takt time for that cell. To determine takt time when there are multiple products running in the same cell, it is necessary to calculate the demand of all products for that cell. It is then a matter of taking the designed daily production rate for each of the individual products, adding them together, and using the total demand for all the products as the designed daily production rate for the cell and dividing that into the schedule time available. This results in one takt time for the cell, which encompasses the demand of all products for that cell.

Workload Balancing and One-Piece Flow

Once a cell takt time has been determined, it is now a matter of comparing several aspects of the process and the takt time in order to design a balanced cell. The operational elements (machine time, man time, and setup time) of each product are examined with relation to takt time. Machine time is compared to takt time in order to determine if the fixed cycle time of any piece of equipment is greater than the takt time. If this is so, action must be taken to change the available time, off load, reduce the cycle time, change processes, add equipment, split demand, etc. If the operation remains greater than takt time, it will need to be balanced with in-process Kanban inventory and/or additional shifts.

Man time is compared to takt time to address two opportunities: (1) autonomation and (2) workload balance. The first opportunity, autonomation, means equipment does not need to be watched in case something goes wrong. Autonomation equipment will automatically shut off when an abnormality is discovered, thereby allowing the operator to do other value-added work. This opportunity is invaluable for increasing productivity and quality. The second opportunity, workload balancing, has to do with examining the individual work elements of each operation and determining if they can be reduced, shifted, resequenced, combined, or eliminated. This effort to balance the workload to takt time is a main enabler for achieving one-piece flow and minimizing manufacturing lead-times.

Setup times are almost always greater than takt time and need to be addressed as part of the cell design process. By comparing setup time to takt time, one has a greater appreciation as to how far setups need to improve in order to create a flexible work environment. The initial stake in the ground is to plan on setting up each high-volume product every day and then to schedule the product mix to run accordingly. If this cannot be accomplished, then plan to run 2 to 3 days' worth at a time and hold the excess inventory until the customer or customer cell asks for it (never allow this to extend past more than a one week's run). It will become very clear, very quickly, why setup reduction is so important, when the supplier cell has to physically

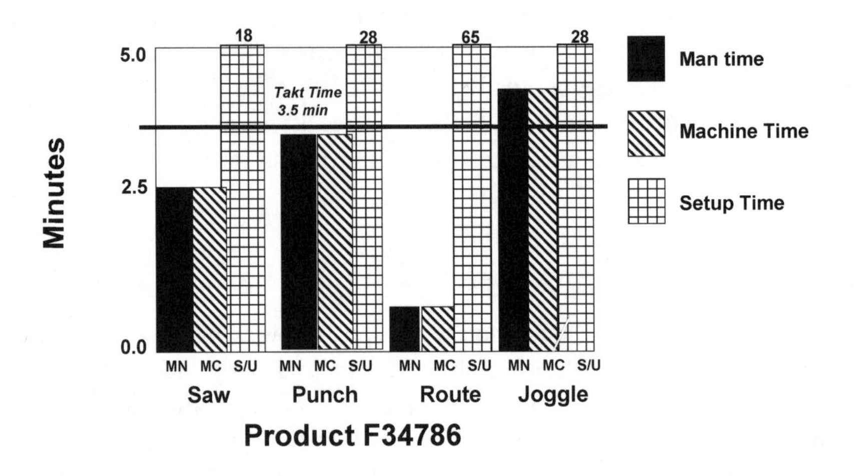

Figure 6.11 Loading Chart

hold the excess inventory until the customer cell asks for it through a Kanban. Once each of these three operational elements is determined for each product, they are compared to the overall takt time of the cell. This information is placed on a loading chart for each individual product SKU (Figure 6.11).

From this point, it is a matter of generating ideas and looking for cell design solutions that will balance the cell workload for all parts and takt time. By reviewing the actual work elements and either improving the operations or shifting the work content, the cell can become more balanced compared to the takt. This is accomplished much more easily in an assembly environment than in a fabrication environment, but it can be done in both.

When the operations are balanced to takt time, it is possible to take advantage of a one-piece flow approach to workflow instead of running in large batch quantities. With one-piece flow, the manufacturing lead-time, level of inventory, and feedback on quality issues are far superior to that of a batch-and-queue system. In a batch-and-queue system, individual pieces are completed at an operation and sit waiting in queue until the entire batch is complete, at which point they are moved to the next operation in sequence and wait in queue for other orders to be completed that arrived there first before moving forward. In the one-piece flow approach, products are passed one piece at a time from operation to operation with a first-in/first-out (FIFO) priority. Product manufacturing lead-times are now only as long as the total of all the takts they had to get through. For example, five operations each with a takt of 1.0 minute require a manufacturing lead-time of five minutes. Another significant benefit to one-piece flow is the impact on quality. There are fewer units in flow to rework or scrap; if there is a defect found, the feedback is almost instantaneous and corrective action is taken on the spot, not several weeks later.

Once we know the cycle time for the process and we know the designed takt time, we can take the known cycle time and divide it by the takt time to determine the maximum staffing requirements for the cell. For instance, the cycle time from the example above was 5.0 minutes. If takt time for that process were 2.5 minutes, then the required staffing would be two operators. Actual head counts will vary with changes in required daily demand, which is why cross-training and operator flexibility are so important in supporting one-piece flow.

Cell Design Criteria

When it comes to designing a cell, there should be established a set of specific design objectives or criteria to be achieved. These criteria are to be the guiding focus for good cell design. The following is a list of general criteria to consider as part of a good cell design:

1. Be sure that material flows in one direction.
2. Reduce material and operator movement.
3. Eliminate storage between operations.
4. Eliminate double and triple handling.
5. Locate parts as close as possible to point of use.
6. Utilize task variation to reduce repetitive motion.
7. Locate all tools and parts within easy reach.
8. Ensure short walking distances.
9. Eliminate all wait time.
10. Keep in mind that vertical storage requires less space than horizontal storage (include Kanban material).
11. Lay out machines and tools by process sequence.
12. Involve operators in the design process (incorporate economies of motion).

Cell Layout

The cell layout is a graphical representation of the operator flow and material flow (Figure 6.12). It depicts the path of the overall material movement through the cell and describes the designed operator sequence and operations.

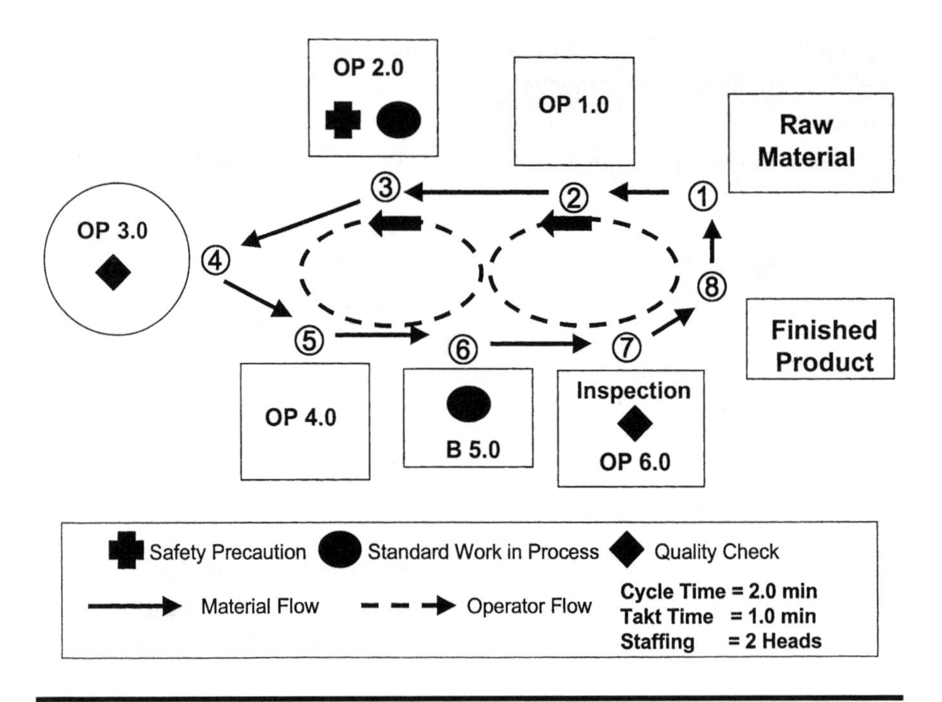

Figure 6.12 Cell Layout

It shows the staffing levels, takt time, cycle time, designed in-process stock levels, and quality and safety checks required by the cell. The cell work layout chart can then be supported at a more detailed level with graphical work instructions for each operation (see Chapter 7). The cell work layout is primarily for training new operators, communicating standard work to management, and driving continuous improvement initiatives.

Kanban Sizing

It is at this point in the cell design process that the control of workflow through Kanban is determined. The number of Kanbans and quantity can be determined in a multitude of different ways. There are several different formulas that can be utilized and which are identified in most operational management textbooks. There are primarily rules of thumb relative to the number of days or weeks of inventory located on the shop floor, and there are simulation model calculations based on the amount of work in process built up in work queues due to process variability. The method of calculation is not that important; just pick one and use it. Most Kanban system implementations fail

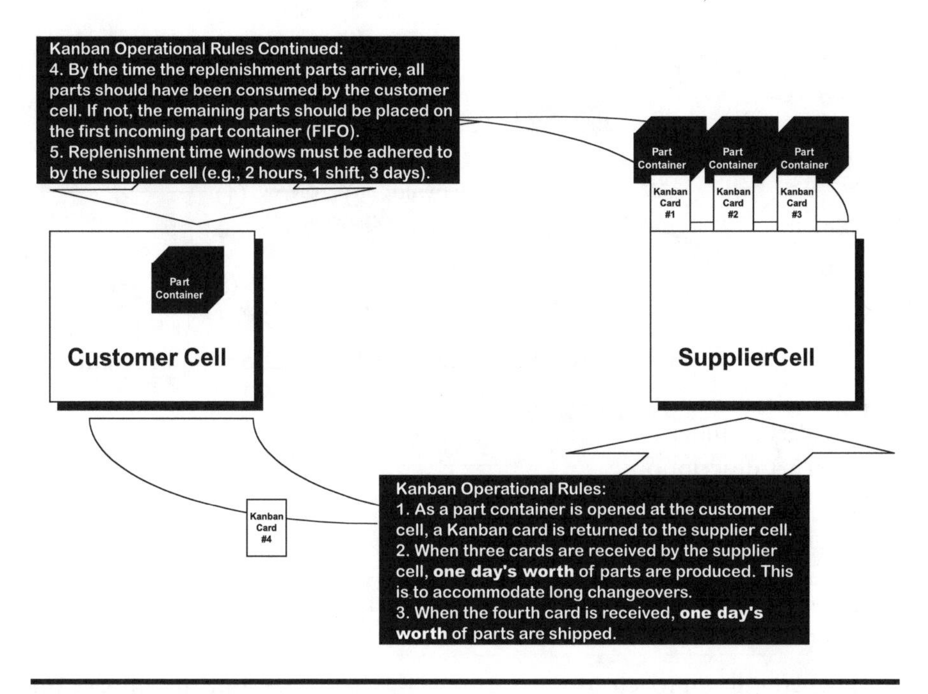

Figure 6.13 Kanban System

because of lack of discipline or lack of training, not because someone used the wrong calculation. That having been said, a simple formula has been included in this section as a point of reference as to how Kanbans could flow between a customer cell and a supplier cell.

Kanban formula:

Step 1.

$$\frac{\text{Designed daily production rate} \times \text{replenishment time (hours)}}{\text{available time}} = \text{Kanban quantity}$$

Step 2.

$$\frac{\text{Kanban quantity}}{\text{lot size}} = \text{\# of cards}$$

Note: Lot size may be required due to weight, size, A,B,C categorization, setup times, common resources, outside suppliers, etc. Replenishment time that is less than one shift would result in a two-bin system. Replenishment time that is greater than one shift would result in a card system (see Figure 6.13). For example:

Step 1.

$$\frac{90 \text{ pieces} \times 15 \text{ hours}}{7.5 \text{ hours}} = 180 \text{ pieces}$$

"A" parts = 1/2 day demand, or 45 pieces

Step 2.

$$180 \text{ pieces} \div 45 \text{ pieces} = 4 \text{ cards}$$

Every Kanban should have the minimum identification requirements:

1. Part number
2. Part description
3. Part quantity
4. Point of supply
5. Point of consumption
6. "One of… cards" (e.g., 1 of 3; 2 of 3; 3 of 3)

The overall approach to determining Kanban sizes and the impact on inventory would include:

1. Gathering the data required for each part number in the cell
2. Utilizing the Kanban calculation to determine the Kanbans in flow
3. Determining the target inventory level based on the Kanban quantity
4. Calculating the designed number of inventory turns

The determination of Kanbans is an important step in the cell design process because Kanbans are the limiting factor for inventory levels (raw material, work in process, finished goods) and are the control element on lead-times. These operational aspects (inventory and lead-time) have a major influence on continuous improvement within a cellular operation.

In his book, *The Just-In-Time Breakthrough: Implementing the New Manufacturing Basics*, Hay described a test for determining if a cell is truly a just-in-time work cell: "The first test is whether the product is flowing one at a time. …The second test to see if a machine cell is truly a JIT cell is whether the machine cell has the flexibility to be operated at different output rates and with different crew sizes."[7] Although I would agree that these two aspects should be evident in order to have a cell, I would hope we have a greater appreciation for just how many other aspects are necessary in order to have a truly successful lean manufacturing cell.

Now that the steps for cell design have been identified and we have greater insight into the impact of material flow through the factory, it is time to address those aspects which lay at the foundation of continuous improvement and provide stability to the cell — namely, the Process Control element.

7 Process Control Element

Process Control focuses on stabilizing the process, institutionalizing the change, and driving continuous improvement activities. The production processes of many manufacturing operations are not in control nor are they performing at the levels necessary to support a lean environment; therefore, there is a need to address these areas as part of the implementation. After a change has been made to a process, it becomes necessary to "lock it down" and maintain it as the new standard for operating; however, after having set the new standard, the performance level should not be limited to that standard, so continuous improvement tools are used to establish a new level of performance. A good management practice to consider implementing would be that of expecting standards to improve twice per year. According to Shingo, in *A Study of the Toyota Production System*, Toyota is extremely rigid in regard to its standards and expects continuous improvement: "The Toyota production system demands that all work be performed within standard times, and shop supervisors are charged with holding workers to those standards. …Shop supervisors are encouraged to feel embarrassed when the same standard operating charts are used for a long time because improvements in the shop operations should be made continuously."[19] This chapter deals with many of the institutionalization aspects of lean manufacturing and describes methods that can be utilized to foster the continuous improvement aspects of a lean manufacturing environment.

This last primary element, Process Control, focuses on a number of lean manufacturing aspects that stabilize the standard methods of working and then continually pursues the setting of new standards for those methods.

This element brings to light several activities that lay the foundation necessary for a company to reach world-class levels of performance, and it is the pursuit of these activities that sets the wheels of continuous improvement in motion, thereby developing processes that are more robust, reliable, and predictable. This chapter highlights six important activities relative to Process Control:

1. Single-minute exchange of dies (SMED)
2. Total productive maintenance (TPM)
3. Poka-yoke (fail safe)
4. 5S (housekeeping)
5. Visual controls
6. Graphic work instructions

Even though these do not address all aspects of the Process Control element, they do provide enough insight for an organization to initiate action on some of the more critical areas.

Single-Minute Exchange of Dies

The implementation of setup reduction is a cornerstone for any lean manufacturing program. The dependency on flexibility (especially in fabrication) is paramount to allowing level production schedules to flow. Following are benefits of the single-minute exchange of dies (SMED):

1. Equipment changeover time measured in increments of less than 10 minutes
2. Minimal loss to throughput time on equipment
3. The ability to run a greater variety of product mix across a given resource
4. Building today only what is needed today

The SMED process is not focused on the reduction of total time spent doing setups, but rather on the pursuit of conducting more setups in the same amount of time. By cutting changeover time in half, a cell can now conduct twice as many setups in the same amount of time. By cutting them in half again, a cell can now conduct four times as many setups in the same amount of time. The primary objective is to build flexibility into the process.

Shigeo Shingo developed SMED as a manufacturing consultant to Japanese companies during the post-World War II era. It took several years for him to perfect the process of setup reduction and design it as a structured set of steps used to deliver incredible capability to organizations that take advantage of this competitive weapon. The process is not terribly difficult, and as much as 75% of the battle has to do with a positive attitude. As Shingo states in *The Sayings of Shigeo Shingo: Key Strategies for Plant Improvement*: "It's the easiest thing in the world to argue logically that something is impossible. Much more difficult is to ask how something might be accomplished, to transcend its difficulties, and to imagine how it might be made possible."[22]

The process has three basic steps: (1) segregate the activities, (2) re-categorize, and (3) reduce or eliminate steps as they are done today. In step one, identify all the activities in the process. Typically, most companies do not really know what their labor force has to go through in order to make a setup. It is not uncommon for an operator to have to travel half a mile (2500 feet) in order to accomplish all the tasks necessary to make a setup. Don't just take my word for it. Walk the entire process sometime. You will be surprised at what you learn. Once all the steps have been documented for the setup process, they need to be segregated into two categories. The first category is that of internal setup — those items that have to be done while the machine is down. The second category is that of external setup — those items that can be done while the machine is running (Figure 7.1).

When the activities have been identified and segregated, the next step is to re-categorize or shift as many activities as possible from internal to external. Typically, between shifting activities from being internal to external

External Set-Up Activities are operations performed while the machine is running (previous or current job)

Internal Set-Up Activities are operations performed while the machine is stopped

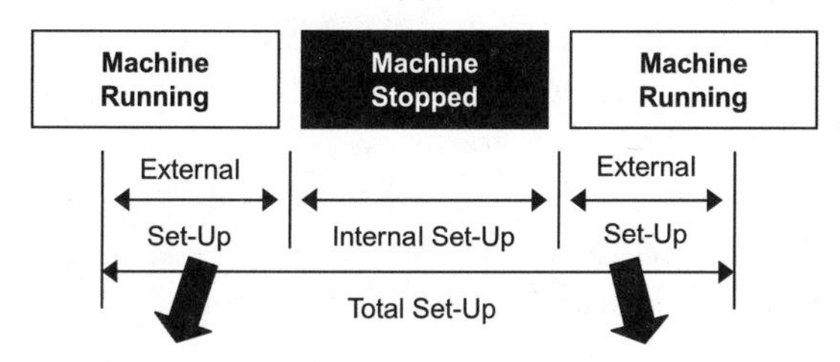

- Retrieve and stage parts, tools for next lot
- Pre-heat, pre-measure, pre-locate
- Verify tool functionality

- Clean and store tools
- Move parts to next operation

Figure 7.1 Identify Internal vs. External Setup

and conducting some good housekeeping practices, changeover time can be reduced by 50%.

Once setup activities have been documented and re-categorized, the last step is to look at simplifying the setup process for both internal and external activities. Investigate standardizing the setup, minimizing the utilization of bolts and adjustments, and utilizing simple one-turn types of attachment methodologies and such techniques as cams, interlocking mechanisms, slotted bolts, secured washers, etc. Strive to make the setup process standard, consistent, repeatable, and one that employees can learn. Too often statements are made about a particular setup process being too highly skilled or too black art or requiring too many years of experience. All of these issues need to be designed out of the setup process. Just as Shingo stated in *A Revolution in Manufacturing: The SMED System*: "It is generally and erroneously believed that the most effective policies for dealing with setups address the problem in terms of skill. Although many companies have setup policies designed to raise the skill level of workers, few have implemented strategies that lower the skill level required by the setup itself."[18] Following the three basic steps, utilizing the techniques mentioned, and having an open mind about the possibilities are all key ingredients to making a SMED program flourish.

Total Productive Maintenance

A second cornerstone in support of a lean manufacturing environment is that of total productive maintenance (TPM). Equipment is integral to any manufacturing environment, and the reliability of equipment in a lean environment is paramount to a truly successful implementation. As inventory levels are reduced, the uptime on machinery becomes even more important. Because there is little inventory to buffer unplanned downtime in a lean environment, when a machine goes down the entire production line goes down; therefore, a formal TPM program is instrumental in supporting a lean manufacturing implementation.

There are three main aspects of a TPM program: preventative maintenance, corrective maintenance, and maintenance prevention. Each one of these components has a different mission and required outcome as part of the TPM program. Each has a significant role to play and is necessary for world-class performance to be sustained.

The first, preventative maintenance, focuses on preventing breakdowns from happening and is by far the most recognized activity relative to TPM.

Preventative maintenance is concerned with the uptime or availability of equipment. The effort here is aimed at performing preventative maintenance actions on equipment in a preplanned/scheduled manner, as opposed to in an unplanned or chaotic manner. Also, the inclusion of operators in this program, specifically to conduct daily maintenance on the equipment and identify abnormalities as they occur, is paramount to successful preventative maintenance. By doing this, the throughput and available capacity on equipment are significantly improved.

Corrective maintenance concentrates on improving repaired equipment. The idea here is that if components from the original equipment keep breaking, why not replace them with something better? Fixing them with an improved component results in longer equipment life and more uptime from the equipment.

Maintenance prevention is an area that most companies neglect and pay very little attention to when designing or purchasing new equipment. Because one of the key ingredients of a successful TPM program is that of daily operator "autonomous maintenance," it is imperative that equipment be easy to maintain on a recurring basis. If the new machinery is difficult to lubricate, if bolts are difficult to tighten, and if it is impossible to check critical fluid levels, then it is very unlikely that operators will be motivated to monitor equipment on a daily basis. The total life-cycle costs on equipment must be examined when procuring new machines, not just the one-off, nonrecurring costs.

In support of TPM as part of a lean manufacturing implementation, the information relative to downtime on equipment is important. Most of the time, if any information is collected at all, it is when equipment has crashed and the cause for the downtime condition is documented. Even though this is good, it provides only a partial picture as to the true throughput loss on equipment. There are in actuality six main reasons, with associated causes, for throughput losses on machinery (Figure 7.2). Shirose identified these losses in his book, *TPM for Workshop Leaders*, and declared them to be negative obstacles to efficiency: "There are two ways to improve equipment efficiency: a positive way and a negative way. …The negative was is by eliminating the obstacles to efficiency — obstacles that in TPM are called the six big losses."[23]

Each of these losses has an impact on the throughput and planned capacity of equipment. Typically breakdown is really the only loss for which we capture information, although all six lead to a reduction in productivity. Breakdown and setup (changeover) have an impact on machinery availability. Minor stoppage and reduced speed have a direct influence on the productivity

Breakdown: Failed function and reduced function

Setup and adjustment: Imprecise and nonstandard measurement

Idling and minor stoppage: sudden disruptions

Reduced speed: actual vs. designed

Quality defects and rework: sporadic and chronic

Startup yield: process instability

Figure 7.2 Total Productive Maintenance: Six Big Losses

of equipment when it is running. Quality and startup yield certainly have an effect on a company's ability to produce, particularly when a portion of valuable capacity is spent on producing poor-quality product.

A technique used to keep the six big losses in check is that of overall equipment effectiveness (OEE), which is measured as a percentage and utilizes information from unplanned downtime, machine cycle time, and process yield to determine which of the six big losses are having the greatest impact, thereby providing insight as to where to focus improvement efforts. An appreciation of these six big losses and how to reduce their effect on equipment resources within the plant will go a long way toward supporting a lean manufacturing program.

Poka-Yoke (Fail Safe)

Human beings will invariably make mistakes. It is not possible to remember everything that has to be done at every step of producing every product with every job. People will make errors; it happens; however, errors are the not same as defects. A defect is what takes place after an error occurs. By sorting good product from defective product at the end of the process, a company cannot hope to achieve a defect-free environment. If, however, errors are caught before they lead to defects, then a defect-free environment becomes possible. This is where the power of Poka-yoke comes into play.

Poka-yoke, another aspect developed by Shingo after World War II, in conjunction with source inspection, was designed to focus on the pursuit of quality at the source and capturing feedback on defects as close as possible to the root cause. In *Zero Quality Control: Sources Inspection and the Poka-Yoke System*, he states: "A Poka-yoke system possesses two functions: it can carry out 100 percent inspections and, if abnormalities occur, it can carry out immediate feedback and action."[21]

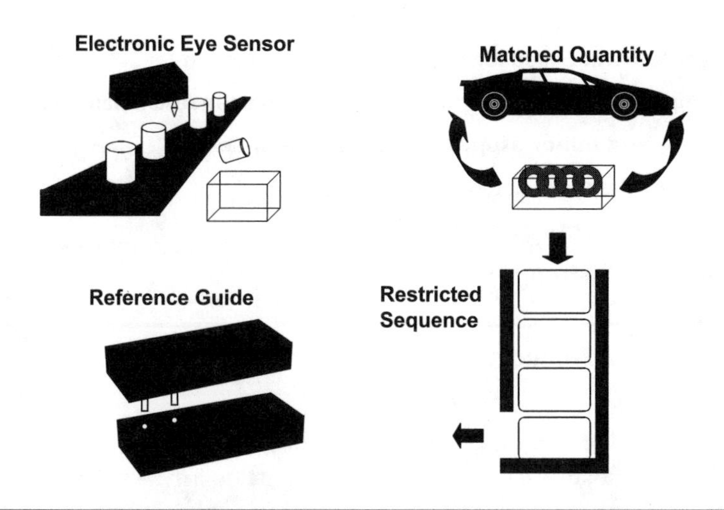

Figure 7.3 Examples of Error-Proof Devices

Poka-yoke, or mistake proofing, is accomplished through the deployment of simple, inexpensive devises designed to catch errors so they do not become defects. These devices are placed in the process to ensure that it is very easy for the operator to do the job correctly or very difficult for the operator to do the job incorrectly. The tools could be physical, mechanical, or electrical (Figure 7.3).

A Poka-yoke could be as simple as a checklist for the operator or technician to ensure that all steps in the process are covered, much in the same manner as pilots going through a pre-flight checklist before taking off. The intent of the Poka-yoke is to stop defects at the source, to provide immediate feedback as to the cause, and to prevent the passing on of defective products to the next customer in the process.

5S (Housekeeping)

Everything has a place and everything in its place! If it does not warrant a label, it does not warrant a place in the area! These are words to live by in a lean manufacturing environment. So, what is so important about housekeeping? According to authors Henderson and Larco (*Lean Transformation: How To Change Your Business into a Lean Enterprise*), it is very important: "Most people underestimate the importance of safety, order, and cleanliness in the workplace. Our former colleagues at Toyota and Honda will tell you that 25 to 30% of all quality defects are directly related to this issue."[9]

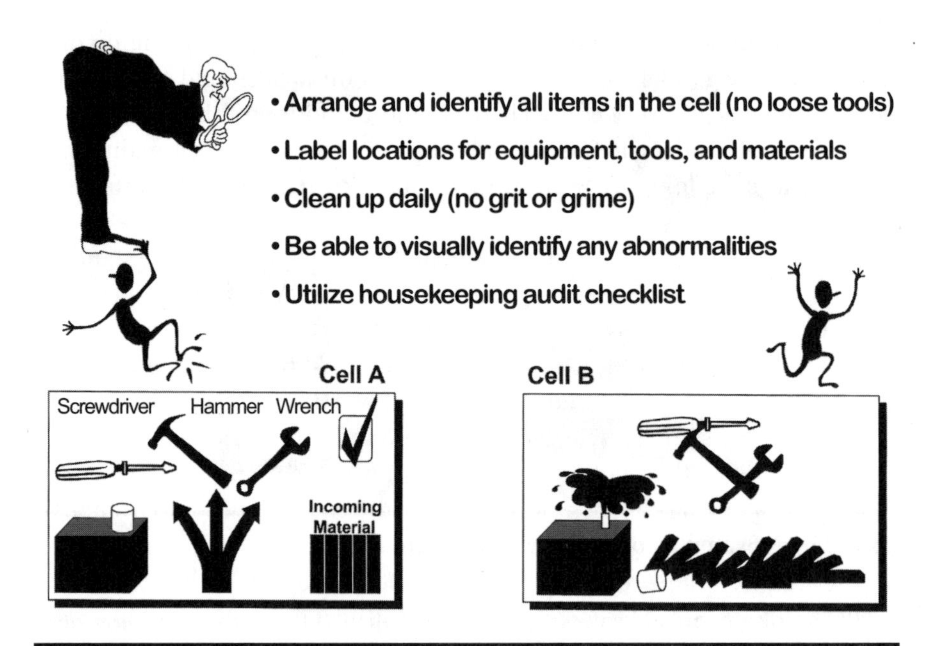

Figure 7.4 5S (Housekeeping)

What is this thing called 5S (Figure 7.4)? For all practical purposes, it represents simple, sound, structured, synchronous, serviceable housekeeping. No, that is not what 5S stands for; rather, the words are all Japanese, coined by Toyota:

1. *Seiri* (sifting)
2. *Seiton* (sorting)
3. *Seiso* (sweeping)
4. *Seiketsu* (standardize)
5. *Shitsuke* (sustain)

The first, *Seiri*, has to do with clearing the area of those items that are not being used on a regular basis (e.g., the next 30 days). It is a matter of sifting through and separating the clutter from the items that are needed to make it easier to work, easier for material to flow, and easier for operators to move, in addition to improving utilization of space.

Seiton deals with identifying and arranging items that belong in the area. These items should all be sorted and labeled as belonging in that area. If the item is not important enough for a label, then it is not important enough to stay in the area. This makes recognition of the proper tooling, resources, materials, etc. extremely visible.

Seiso has to do with maintaining order by sweeping and picking up on a regular basis (e.g., daily, bi-weekly). A production area should be neat and clean at the end of every shift. There should be nothing missing or out of place. All tools and materials should be accounted for. A well-maintained area should be able to accomplish this using less than 2% (10 minutes) of the daily scheduled shift time.

Seiketsu is concerned with management discipline to enforce the standard activity. If the housekeeping activity does not become institutionalized within the operation, the area will not stay clean and employees will revert back to the old ways very quickly. A regular, formal audit with quantitative and qualitative expectations should be conducted and scores posted for areas of responsibility. Assigned areas of the floor are important, because if everybody has responsibility, then nobody has responsibility.

Shitsuke is management's responsibility to reinforce the importance of housekeeping and to demonstrate leadership by follow-through and walking the talk. People will pay attention more to what management does than what they say. Proclaim that housekeeping is important, clarify expectations, walk the shop floor, reward those who are performing, and constructively discipline those who are not.

Visual Controls

The area of visual controls encompasses such concepts as line-of-site management, or the ability to walk onto the shop floor and in a matter of minutes know the status of the operation, what might be abnormal, how the material is flowing, what job is in work and what job is next to go in work. It also includes the concept of signage, which means that everything is displayed, marked, documented, and reported, so much so that any individual off the street could walk into the factory and give a plant tour.

A key aspect of visual control is that of shopfloor performance measurement, accomplished through the display of a handful of measures (three to five) on the shop floor for everyone to see and understand. As was stated in Chapter 4, these are to be measures that are created, owned, monitored, controlled, and understood by those in the area. If a measure is created in another area, then brought to the shop floor and posted in another area, it is very unlikely that people working in that area will really know what it means. Worse yet, they could not explain how their job performance relates to that measurement. It is important for individuals to understand whether their areas are performing to plan, it is important for them to record how

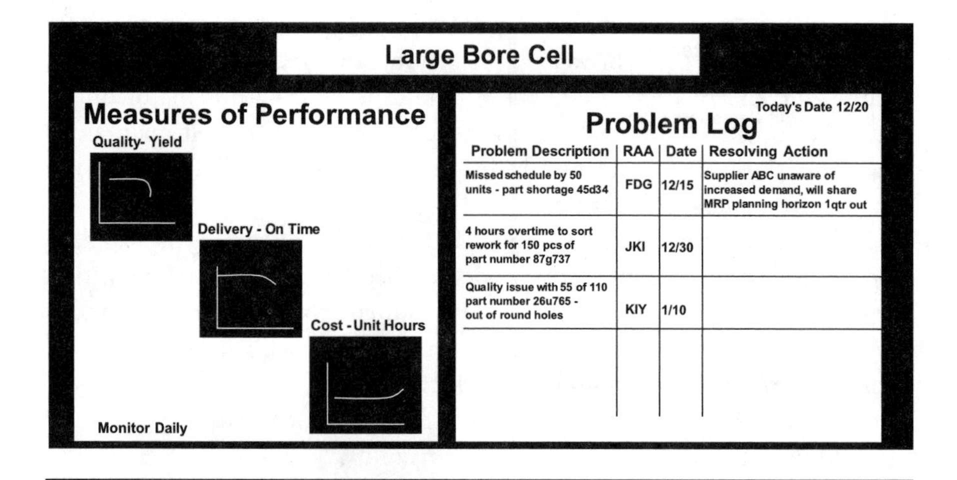

Figure 7.5 Communication Board

they are doing according to that plan, it is important for them to take responsibility for that performance, and it is imperative that they know how to improve that performance. In addition, they need a way to highlight problems in order to receive prompt support and corrective action.

The utilization of a visual control board or communication board (Figure 7.5) will provide the means to display performance status and communicate problems. Typically, the communication board is divided into two halves. One half contains the shopfloor measures of performance (e.g., schedule adherence, quality, cycle time, etc.). The other half contains a problem section, where the operators can document problems they are having. These problems are reviewed on a daily basis, actions assigned, resolution dates committed, and mitigating actions recorded. This provides visibility to shopfloor problems that are otherwise hidden or placed on a list to be resolved someday. The importance of visual controls is how they make improvement activities, issues, performance status, problems, and operational rules visible.

Graphic Work Instructions

To consistently convey how a job is to be performed according to documented standard work sheets, the message needs to be communicated in an easily recognizable format. Text-based work instructions are probably the least attractive means of accomplishing this task and yet are by far the most widely utilized, probably because this has been the easiest way to bring information

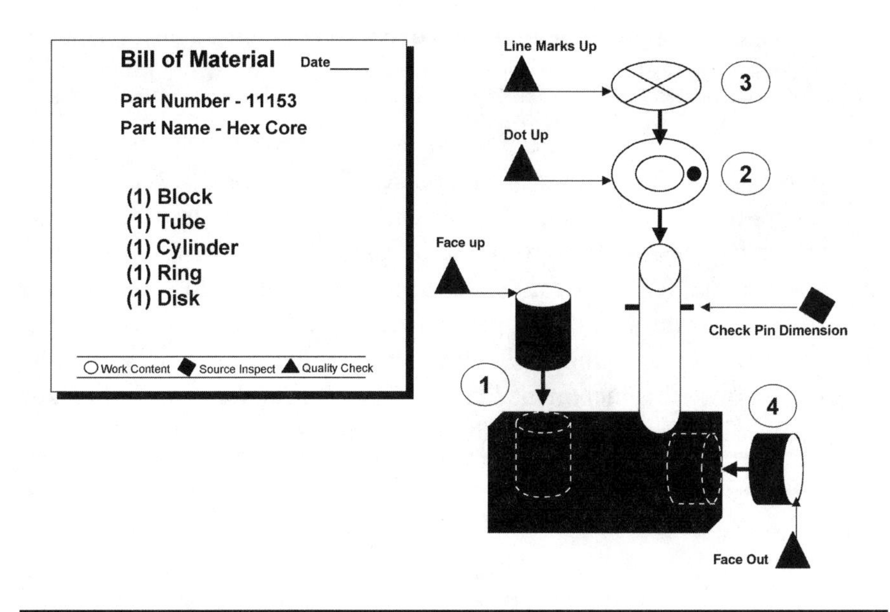

Figure 7.6 Graphic Work Instructions

to the shop floor. The problem with text is that it is very dependent not only on an individual's ability to learn from a written format but also on an individual's ability to accurately describe actions as part of a series of motions, not to mention the cross-cultural language barriers that can exist within the plant or when communicating globally regarding products or production methodologies.

In the past, CAD drawings and blueprints were the only means of graphically depicting work and were very time consuming to update and maintain; however, with the advent of digital cameras, video recorders, and presentation software, there is no excuse for not providing graphic instructions in the shop area. Graphic-based work instructions are a far more effective means of communication than simply text (Figure 7.6). The information can be captured quickly through a digital camera and manipulated with software to add color-coded legends that identify work content by operation, quality checks, special notes, etc. The beauty of color is that it can transcend language barriers. If there is a concern over employees who are color blind, make use of standard symbols. A green circle could represent work content; a yellow triangle, a quality check, etc. Exploded views, particularly of assembly operations, are of tremendous benefit, but they do require specific software applications.

Each picture or slide can represent an operation or depict a bill of material for that operation with a date, revision, and signature block for configuration control. When there is an improvement to the process or the introduction of a new part, the old graphic can be pulled and replaced with a new one in as little as 30 minutes. The days of a manufacturing engineer having to spend several days trying to maintain and update work instructions are over.

The deployment of all five of these primary elements of lean manufacturing is essential for most companies to achieve true world-class levels of performance. It is through the proper sequencing and timely implementation of these elements that a company can achieve that illustrious level of performance. But, once this incremental change in performance is achieved, how is it sustained? The next chapter will briefly touch on this issue.

8 Sustaining the Change

Now comes the answer to the great mystery of how to keep things the way you want them. The answer is … you don't! You do not want this process to be static. You most assuredly do not want it to fall back, but you do not want it to stay the same either. When companies stay the same, they fall behind. Change is a constant; therefore, constantly looking for new ways to improve the business is the name of the game. If companies are not improving, at least at the rate of inflation, then they are losing, and having to face pricing pressures from the market just compounds the seriousness of the situation. Companies need to constantly push themselves to challenge the *status quo* by performing better today than yesterday. So how is this achieved? First of all, there is a need to institutionalize changes that have been made to this point by doing the following:

1. Develop and deploy housekeeping audit checklists (i.e., 5S) and have the management discipline to review them at least once a month.
2. All operational work is standardized, displayed, utilized as a basis for continuous improvement activities, and improved twice per year.
3. Each manufacturing cell team is expected to conduct a Kaizen event every quarter.
4. Shopfloor performance measures are updated by the cell team daily.
5. Problem boards are reviewed at the end of every day.
6. Cell teams hold standup meetings every day to review progress and issues.
7. Actual setup times are recorded with each changeover.
8. Cross-training skill matrices are updated quarterly.
9. Equipment is cleaned and checked on a daily basis.
10. Customers and suppliers are visited by the cell team each quarter.

After the change has truly been institutionalized, a focus on growing the business through new products or markets and production capability is next. The cell team accomplishes this by:

1. Looking at the producibility of its existing product base.
2. Determining how lean the current product design is and identifying where opportunities exist to take additional waste out of the design.
3. Utilizing failure mode and effects analysis (FMEA) to improve the robustness of the manufacturing process and minimize risk of poor-quality output.
4. Looking for opportunities to pilot new production.
5. Looking for potential improvements within the supplier base.
6. Understanding cost, quality, and schedule issues with suppliers and helping them to identify and eliminate waste from their processes.

When you have reached this level of capability within your manufacturing organization, you are definitely well ahead of the pack and have reached a level very few have actually achieved; however, do not stop there. Remember that lean manufacturing is the continual pursuit of improvement and it is a journey that never ends.

PUTTING IT ALL TOGETHER

9 Setting the Stage

This section of the book shows how to design and deploy a holistic lean manufacturing program. The following chapters contain a fictitious business story in which many of the tools and techniques described in this book are utilized. Picture yourself in these chapters as the new Director of Lean Manufacturing for this company, and follow along in the story to learn how to design, develop, and deploy your own lean manufacturing program. The story describes a current business situation and demonstrates use of the tools via completed templates to show how a lean manufacturing program is developed and managed. Take the time to read through the story and understand the templates, as learning by doing is one of the best ways to retain knowledge. Obviously, one cannot instantly grasp all of the potential issues that need to be addressed when deploying a lean manufacturing program; however, I have tried to include many of the most common issues that have arisen over the years during my implementations. I hope you find this story both worthwhile and entertaining.

Setting

Regal, Inc., in Cincinnati, OH, has been in business since 1945. They started out as a small, subcontract supplier to the heavy industrial capital-goods market and began to grow when they picked up small, niche-market contracts for the machining of specialty bearings, housings, and pistons. The housing and piston work came as an offload opportunity when the primary supplier was overloaded. Regal did good, high-quality work and had excellent response time.

Over the years, Regal was able to expand the business through niche growth in the marketplace and positioned themselves as the "go to" player with the original equipment manufacturers (OEMs) when it came time to design new products. Their reputation and expertise opened many doors. These opportunities led to further growth in the market and growth within the business. The company relocated to a 500,000-square-foot facility across town in 1980 and invested significantly in additional capital equipment and vertical integration to meet the demands of the increased volume. They had been achieving revenue increases of about 15% each year for the last 5 years and were enjoying operating profits of 20%, with no end in sight.

So, everything was looking good until one day one of their first OEM customers, who had been with them since the 1950s, announced the rollout of a new piece of equipment which Regal knew nothing about. When the company asked the OEM why Regal had not been utilized for this new design, they were told that a different supplier had been more responsive by offering shorter lead-times, more consistent delivery performance, and more reliable product quality, plus they promised a cost reduction of 5% each year for the life of the contract and no hassles about delivering products to the OEM plants as needed on a daily basis. Regal's management viewed this as only a minor setback, until a second long-time OEM canceled an existing contract, paid the penalty for doing so, and went with a competitor for the same reasons.

With this additional loss in business, in order to maintain their 20% operating profit, management believed a reduction in work force was the next logical alternative. On the day they were to make the final decision, a recruiter called the Vice President of Operations at Regal and explained that he had the resumé of an individual that the company might be interested in seeing. Because the vice president and the recruiter were long-time fishing buddies, he agreed to review the resumé and had it faxed over.

Main Cast

President: Brian Stevens
Vice President of Operations: David Brice
Vice President of Sales and Customer Service: Barbara Stearn
Vice President of Product Development: Samuel Button
Director of Human Resources: Heather Dale
Controller: Joseph Billings
Director of Information Systems: Paula Wright

Director of Material Management: Steve Phelps
Plant Manager: Richard Johnson
Manager of Production Engineering: Carl Withers
First-Line Supervisor: Jake Holden
Director of Lean Manufacturing (newly hired): Robert James

Diagnostic Information

Sales 1997 — $63 M
Sales 1998 — $73 M
Sales 1999 — $83 M
Sales 2000 — $65 M (projected)

Profit 1997 — $12.0 M
Profit 1998 — $14.6 M
Profit 1999 — $17.4 M
Profit 2000 — $13.0 M (projected)

Headcount 1997 — 420
Headcount 1998 — 487
Headcount 1999 — 553
Headcount 2000 — 433 (projected)

The Interview

The next day, Robert James arrived at Regal, Inc., for an 8:00 a.m. interview. He was ushered into the building and deposited at the office of David Brice, the Vice President of Operations. At about 8:20, David rushed into the office, out of breath, and introduced himself to Robert.

"Good morning! This place is a mad house," exclaimed David. "I don't know how we could have lost that business, but we will just have to suck it up and work harder to make it happen, I guess."

"What business is that?" inquired Robert.

"Oh, a long-time customer of ours, B&D Industries, has decided to design and develop its latest product without involving us."

"Why did they do that?" asked Robert.

"Oh, they made some excuse about us not being responsive to their needs and our continuing to raise prices on them," replied David.

"Were they correct?"

"Not as far as I'm concerned. You see, we have been doing business with them for nearly 40 years, and just because some new player has come into the marketplace, making impossible claims about reducing prices year after year and responding to their schedule needs on a daily basis, they have decided to change their loyalties."

"Can Regal meet those identified performance requirements?"

"No way! If we made those kinds of outrageous commitments, we would lose our shirt! Our quality would suffer, and the rest of our customer base would be impacted."

"If the marketplace is asking for those kinds of requirements, and your competition is demonstrating the ability to satisfy those requirements, don't you think you may have more than a short-term profitability problem?" queried Robert.

"I am not so sure that this competitor, Blue Iron, can actually deliver what they say. Their operation is located in the southwest, which is not as geographically close to B&D as we are, and, besides, the delivery responsiveness that Blue Iron is claiming is unheard of in our industry," explained David.

"Well," Robert said, "my brother-in-law happens to work at Regional Consolidated, which is a major customer of Blue Iron, and they *do* deliver on those expectations. They do reduce prices each year through cost reductions and they do respond to scheduled needs of the customer."

"Do they really?"

"Yes, they do! Let me ask you, do you get out much to talk with customers or do you compare Regal's products to the competition's?" asked Robert.

"No," said David, rather sheepishly. "We don't get out much at all. So much of our time is spent keeping the operation running that there is no time to get out and see customers or compare products."

"Let me ask you this. What percentage of your current sales base is made up of new products? I mean products that have been introduced within the last three years," asked Robert.

"We have targeted about 5 to 10%. We are currently at about 5."

"How have you been able to sustain the growth you've had without introducing new products at a higher rate?"

"Most of the sales growth, in the last couple of years, has come from price increases on our current products, because our unit volume has been flat."

"Would you expect those existing markets to have requirements similar to B&D in the near future?" asked Robert.

"I don't know. I guess I never really thought about it."

"Well, if Regal, Inc., has any indication that this could be the new required level of performance in the marketplace, I would recommend that you look at a different way to align your operations to perform at that new level," Robert offered.

"Do you know of a way to do this?"

"That's why I'm here. So, let's talk…"

10 How It Begins

When David introduced Robert to Regal's top management team as the new Director of Lean Manufacturing, there was a fair amount of eye rolling and blank stares in the room. David explained to the group that, due to the recent developments with B&D, there may be the need to re-examine how they currently conduct business. "There is an indication," said David, stretching the truth a little, "that this may be only the beginning of a long wave of competitive erosion of our customer base. We need to revisit how we are currently conducting business before we just eliminate heads from the payroll."

Brian Stevens, President of Regal, asked, "Why do we need to revisit our current mode of operations? If we just get some of the excess heads off the books and make everyone aware they need to work harder, we should be all right. We can weather this storm. Besides, this competitor won't be able to deliver on these promises, and B&D will come back to us, hat in hand, within the next six months."

"I'm afraid that's not true," replied Robert. "I know about this company through several of its current customers, and Blue Iron *does* deliver on their promises. They do quite well in their niche markets and are beginning to expand into additional areas, Regal's being one of them. They appear to target markets that have growth opportunities coming through new product developments. They align with customers who are looking to attain the next level of performance and who are disenchanted with their current supply base of mature, slow-moving companies."

"What level of performance are we talking about?" asked Barbara Stearn, Vice President of Sales and Customer Service.

"The benchmark for many companies striving for world-class levels of performance would be 50+ inventory turns per year, same-day delivery on customer orders, manufacturing lead-times of one week maximum, in-process quality levels approaching 99% roll-through yield, and annualized cost reductions of 5 to 10% each and every year," said Robert.

"Those performance levels are unheard of in our industry!" barked Richard Johnson, Regal's Plant Manager. "Not one of our customers is expecting us to achieve those levels of performance."

"I'm afraid that customers have a funny way of deciding what is and is not an acceptable level of performance," said Robert. "You see, the performance target is constantly changing, and if one of your customers hears about a competitor who is achieving such levels of performance, that now becomes the new standard for that customer. Think about it yourself, as a consumer. Ten years ago, when you wanted new or replacement parts for your car or some consumer electronics gizmo, you went to the retail outlet, told them what you wanted, and hoped that they carried it in stock. If they didn't, then you were placed on backorder and the part may have shown up 4 to 6 weeks later. Today, you search the Internet for what you want and order it, and it arrives at your door in many cases the next day. Ten years ago, most consumers would have never dreamed of that kind of responsiveness, but they are coming to expect it today, just as B&D is now demanding new levels of performance from its supply base."

"But B&D is only one of many customers we have. Surely they won't all demand that level of performance, will they?" asked Barbara.

"I don't know, Barbara. Have you asked them lately? Have we inquired about what performance level they need or are receiving from the competition? Do we know how we stack up? Are we leading or lagging? As head of sales and customer service, do you have any information relative to this?" inquired Robert.

"We keep some information in our database as to the competition, but it is gathered only when we introduce a new product line, which has been a while," stated Samuel Button, Vice President of Product Development. "In addition…"

After about an hour of discussion among the management team, they finally reached the general consensus that Regal, Inc., was not really prepared to compete in the marketplace of the future. They all agreed it was a good idea to bring Robert on board to let him guide their operation down the path to becoming a lean manufacturer.

11 The Game Plan

The project team assigned to design, develop, and deploy this lean manufacturing program met the following Monday, August 2. The team consisted of seven full-time, dedicated employees, including the team leader, Robert. The team consisted of Heather Dale, from Human Resources; Joseph Billings, the controller; Paula Wright, from Information Systems; Steve Phelps, who represented materials management; Richard Johnson, who represented plant management; and Carl Withers, from Production Engineering. They spent a significant amount of time that morning discussing why they were together, why there was a need for this team, why were they selected, what they were expected to accomplish, etc. Richard spent much of the morning explaining who he was, why he was there, why they were there, and why this was an extremely important program for the future of the organization.

After about four hours of discussion, debating, venting, and clarifying, they eventually became comfortable about the project and its objectives. They spent time writing out a project charter (Figure 11.1) to clarify their understanding with executive management in regard to the overall scope and objectives for the project. They identified potential risks, issues, and assumptions about the project. Through Robert's facilitation, the team identified specific goals for the lean manufacturing program, developed an overall rolling-wave milestone plan (Figure 11.2) that covered the project duration, and assigned subject matter experts aligned with the Five Primary Elements as follows:

- Organization Element — Heather Dale
- Logistics Element — Steve Phelps
- Process Control Element — Carl Withers
- Manufacturing Flow Element — Richard Johnson
- Metrics Element — Joseph Billings

Regal, Inc.	**Project Charter**	Form: 014
Title:	**Lean Manufacturing Program**	**21**
Purpose:	Design, develop, and implement a lean manufacturing environment by focusing on the value stream for in-house manufacturing and material flow.	
Objective:	(1) Autonomous production units. (2) Self-directed work teams; reliable and predictable demand management. (3) Knowledge transfer of lean manufacturing techniques. (4) Mobilize cross-functional project team. (5) Facility layout and product performance responsibility aligned by product grouping. (6) Assess and select cell team leaders.	
Outcomes:	(1) Improved delivery performance from 56% to 98% to CRSD. (2) Manufacturing lead-time of less than 1 week for all product groupings. (3) Inventory turns (RM, WIP, FG) of 35. (4) Direct labor productivity improvement of 25% on runner products. (5) A 50% reduction in all identified NVA activities.	
Project Owner:	Brian Stevens	
Team Leader:	Robert James	

Figure 11.1 Project Charter

Because the team had developed this plan together, they had a common understanding of where they were going, what they were going to accomplish, and what success looked like when they got there.

On August 9, Robert had the team meet with executive management to demonstrate their understanding of the assignment, to assure clarification of project direction and project duration, and to establish goal alignment. After they received executive management's permission to move forward, the project team produced a 10- to 12-slide presentation for executive management to deliver to the organization. It contained an overall story line explaining:

1. The current state of the business
2. Why a lean manufacturing project team had been assembled
3. The project team's charter
4. The overall schedule (milestone plan)
5. Management's commitment to keep everyone informed as to project progress
6. How everyone would fit into the operation when it was designed

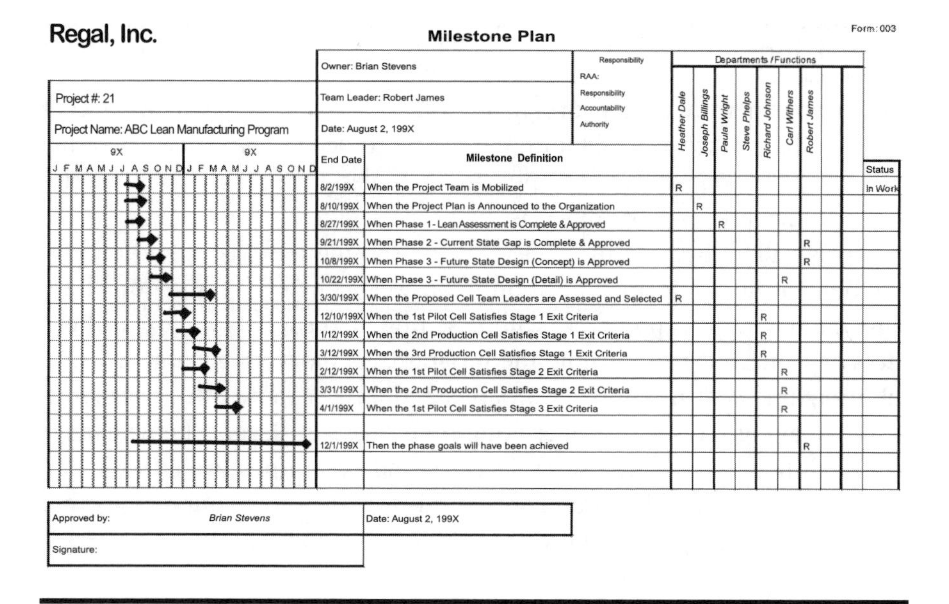

Regal, Inc. **Milestone Plan** Form: 003

Owner: Brian Stevens

Project #: 21 | Team Leader: Robert James

Project Name: ABC Lean Manufacturing Program | Date: August 2, 199X

Responsibility RAA: Responsibility Accountability Authority

9X J F M A M J J A S O N D | 9X J F M A M J J A S O N D

End Date	Milestone Definition	Heather Dale	Joseph Billings	Paula Wright	Steve Phelps	Richard Johnson	Carl Withers	Robert James			Status
8/2/199X	When the Project Team is Mobilized	R									In Work
8/10/199X	When the Project Plan is Announced to the Organization		R								
8/27/199X	When Phase 1 - Lean Assessment is Complete & Approved			R							
9/21/199X	When Phase 2 - Current State Gap is Complete & Approved							R			
10/8/199X	When Phase 3 - Future State Design (Concept) is Approved							R			
10/22/199X	When Phase 3 - Future State Design (Detail) is Approved						R				
3/30/199X	When the Proposed Cell Team Leaders are Assessed and Selected	R									
12/10/199X	When the 1st Pilot Cell Satisfies Stage 1 Exit Criteria					R					
1/12/199X	When the 2nd Production Cell Satisfies Stage 1 Exit Criteria					R					
3/12/199X	When the 3rd Production Cell Satisfies Stage 1 Exit Criteria					R					
2/12/199X	When the 1st Pilot Cell Satisfies Stage 2 Exit Criteria						R				
3/31/199X	When the 2nd Production Cell Satisfies Stage 2 Exit Criteria						R				
4/1/199X	When the 1st Pilot Cell Satisfies Stage 3 Exit Criteria						R				
12/1/199X	Then the phase goals will have been achieved							R			

Approved by: *Brian Stevens* | Date: August 2, 199X

Signature:

Figure 11.2 Milestone Plan

When Brian delivered the message to the organization, through a town hall meeting, he opened by saying, "We at Regal have enjoyed many years of success, and we wish to continue in that same tradition of success. The way we have conducted business up to this point has brought us all a great deal of benefit. We have achieved great success in growing our operation and should be proud of our accomplishments. However, in order for us to continue growing our company, we need to look at conducting business in a different manner. Competition is getting tougher, and it would appear there are a number of companies nipping at our heels and looking to take our customers away. We cannot continue to survive without a customer base, and our customer base is becoming more and more demanding.

"So, as our customer's requirements change, so too do we need to change. Therefore, in light of this situation, we have assembled a cross-functional team staffed with some of our best players, who will be working full time for the next 6 to 9 months on designing and implementing a new manufacturing operation. They will be coming to you for information, asking for your input, and seeking your help. I would ask that you provide them with honest, factual information and when asked for your opinion to respond openly. As part of the analysis and design process, they will be coming to you for concurrence and feedback regarding the design. Your inputs are important. Within the next three months, we will begin to implement this program and will again be soliciting your ideas; however, during implementation, you will be the major players in the process because you will be involved in the actual design and arrangement of your work areas. You will have a say as to what goes where and how the work will flow. We will be guided by some new operating principles, but you will have an opportunity to design out many of the wastes that are currently part of your existing processes.

"This will all become more clear as the coming months unfold and we will be in a more informed position to answer many of the questions I am sure you have at this point. We will be setting up a suggestion box for both ideas and questions concerning the program. As the team gets further into the program, we can report on progress and answer more of the questions as we go along. This is a very exciting time for us at Regal. I know change can be difficult and a little scary, but if we all keep a positive attitude and open mind as to what develops, I am confident we will come out on the other side a much stronger and more capable organization for our customers. I thank you in advance for your support."

12 Lean Assessment

After Brian's announcement to the organization, the team was ready to begin phase one — Lean Assessment. The project team set up their war room, went through three days of intense lean manufacturing training on the Five Primary Elements with Robert, and began the task of assessing the overall level of leanness of the operation (Figures 12.1 to 12.4).

By August 19, the team had gathered the lean gap analysis information by production process and loaded it into the project database. They were now ready to begin documenting operational performance data by both product group and process (Figure 12.5).

Before they began collecting the performance data, Paula made the comment to the team, "I believe much of the data we need are contained within our business system."

Robert stated, "Even though that may be true, Paula, I would encouraged the team to go to *gemba* [the Japanese word for work site] to retrieve the data. Even though much of the identified data could reside in the system, it may not be accurate, and this initiative needs to be a very hands-on program. In addition, it is important for us to be seen on the shop floor, talking with the operators and gathering their insight. They are going to be very skeptical at this point, and we need to be keenly aware of their concerns. We will need to use the system-generated data, but just not as the first source at this time."

The team created a baseline template for the data collection, broke into pairs, and went to the shop floor to learn about the current manufacturing processes. They already had an idea about the current weaknesses in the operation based on the lean assessment scoring, which was completed earlier. The team segregated the shop by assembly, fabrication, and product groups. Heather and Carl took assembly, Richard and Robert took fabrication, and Joseph and Steve took product groups.

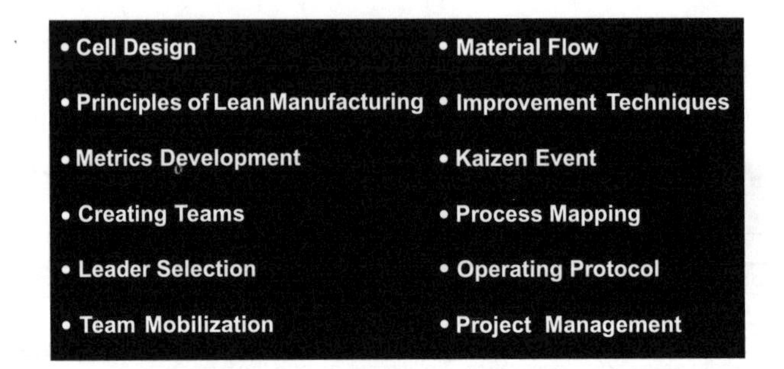

Figure 12.1 Continuous Training

Lean Manufacturing Self Assessment	0	1	2	3	4
Manufacturing Flow					
1. Does material flow "one way" throughout your plant?	x				
2. Is your mfg. process designed to have operators touch material only one time?		x			
3. Do you have production areas aligned to end customer products?		x			
4. Are work stations designed to meet daily customer demand?			x		
5. Is product passed between operators one piece at a time?	x				
Organization					
1. Are cell/product line leaders held accountable for end product performance results?	x				
2. Do you utilize cross-functional teams on the shop floor?		x			
3. Are the roles/responsibilities for all team members defined?		x			
4. Do operators know all steps in the manufacturing process for their area?			x		
5. Are support resources located on the shop floor?			x		
Logistics					
1. Do production areas build to customer demand?			x		
2. Is material pulled between stations?		x			
3. Does the shop floor produce to a daily build schedule?	x				
4. Is material replenished to an A,B,C segregation?	x				
5. Are shop floor operations rules documented and understood?		x			
Metrics					
1. Are performance measures visible and current on the shop floor?	x				
2. Is your schedule adherence 100% on-time?		x			
3. Is your manufacturing leadtime less than 1 day?			x		
4. Is your shop floor targeted performance continually improving?			x		
5. Do your shop floor operators own and report their performance data?		x			
Process Control					
1. Are changeover times on your bottleneck resources less than 10 minutes?		x			
2. Do you have a formal continuous improvement program?		x			
3. Is the reponse time, to defects found in the production process, less than 15 minutes?			x		
4. Do operators have the authority to "stop the line" when defects are discovered?			x		
5. Do you have a philosophy "everything has a place and everything in its place?	x				
Sub Total	7	10	8	0	0

Figure 12.2 Lean Manufacturing Benchmark

When asked by Heather what he thought about all this, Carl answered, "I don't know. It seems like a sound approach on paper, but that is only paper. What happens in implementation, now that's reality. I am not quite sure yet how we design out our current operational problems and develop a system that is responsive to these new levels of performance. What do you think?"

7	10	8	0	0
x0	x1	x2	x3	x4
0	10	16	0	0
	Grand Total			26

Rating Key
0 = Practice not utilized, 0 % occurance, "never been done"
1 = Practiced in some areas, 25% occurance, "just started"
2 = Practiced in many areas, 50% occurance, "some benefit achieved"
3 = Practiced in most areas, 75% occurance, "rolling out across the company"
4 = Practiced in all areas, 100% occurance, "long established practice & institutionalized"

Scoring Key
81-100 (Keep up the tremendous work)
61-80 (You are on the right track)
41-60 (You demonstrate understanding of lean, but may need guidance to reach the next level)
21-40 (You need significant assistance to become a lean operation)
0-20 (You need to change your way of doing business)

Figure 12.3 Lean Manufacturing Benchmark: Scoring

		Cell Implementation Audit				
						Scoring Key
Company Name - Cincinnati, OH			Scoring			1 - Partially Evident, 2 - Mostly Evident, 3 - Completely Evident
Date - August 14th, 199X						
Area	Item Num	Element	1	2	3	Comments
Manufacturing Flow	1	Cell name & boundary posted	1			Cell name was hardly visible; however, boundaries were well marked on the shop floor. A larger (poster-size) cell name should be utilized.
	2	POU areas clearly defined, marked and utilized			3	Areas identified very well by color coding.
	3	Visible display of material flow through the cell	1			No graphical process map ofoverall material flow for continuous improvement and communication, however the process was well understood by the implementation leader (Vega).
	4	Regular utilization of housekeeping audit checklist	1			Checklist is being developed and a form of scoring is being posted.
	5	Two bin Kanban utilization on all part numbers			3	It was very evident on all part numbers viewed. PCB subassemblies are using Kanban; however, not at the planned levels.
	6	Workable work elements identified & stored within the cell area			3	A workable work checklist is being utilized.
	7	Documented process for verifying workable work		2		A workable work process was supposed to have recently been documented; however, it still requires additional development.
	8	Visible and configuration controlled graphic work instructions			3	Excellent!! Configuration controlled, dated, signed.
	9	Kanban sizing calculated and implemented	1			Rule of thumb determination. C parts 1 week & A parts 1/2 day. Not really documented for repeatablity.
	10	Equipment rearrangement complete			3	Complete.
	11	New equipment purchased, installed, operational	1			Scissor lift CER approved and PO placed.
	12	Utilities connected and operational			3	Full functional per cell implementation leader.

Figure 12.4 Cell Audit

"I am concerned about how the people are going to perceive the lean manufacturing program," said Heather. "I mean, we want to involve them and solicit their input, but I just don't know how they are going to buy in to the changes. It seems to me that, to engage them in the process, we need to

- Space (sq. ft.)
- WIP level ($ or equivalent)
- Travel distance (parts and people)
- Manufacturing lead-time (units)
- DTD lead-time (days)
- Output/person/unit (pc/minute)
- Efficiency (%)
- Changeover time (minutes)
- Staff level (heads)

Results reflected by process, by product

- Unplanned downtime (minutes)
- Scheduled time (hours)
- Actual time (hours)
- Planned output (units)
- Actual output (units)
- Planned mfg. cycle time (minutes)
- Actual mfg. cycle time (minutes)
- # of units reworked
- # of units defective
- Employee turnover (%)
- Employee absences (%)
- Annual output volume (units)

Figure 12.5 Lean Assessment Data Collection Items

find out what would motivate them to change. Show them where they fit into the program."

"I think you have a good point there. When we get a chance, we need to talk with Robert about those issues," said Carl.

As Richard and Robert made there way to the back shops, where the fabrication operations were located, a first-line supervisor named Jake approached Richard and asked, "Am I going to have a job when this is over, Mr. Johnson? Because I have a cousin over in Louisville who went through one of these 'lean things' and they laid off nearly half the plant and outsourced almost all the work to somewhere else."

Richard reassured him by saying, "Jake, you do not have to worry about losing your job as a result of this lean program. When all is said and done, your job may have changed or you may be doing a different job, but you won't be eliminated from the payroll, unless, of course, you do not want to work in the new lean manufacturing environment. You see, the thing is if we don't do something like this now there is a good chance I will need to send people out the door later, and I don't want to do that."

"I understand," said Jake.

As Steve and Joseph made their way to the shipping area to ask the packers questions pertaining to the handling times of SKUs, Joseph made the statement, "This program is really going to play havoc with my overhead absorption numbers. All the individual department allocations are measured by each operation's hours produced per day. If we start changing the focus to

Company: Regal Inc.	Product Family: Razor - High Performance Pistons

Type	Current ABC Units	Current ABC Dollars	Current Gross Margin %	Current ABC Trend %	Total Market Units	Total Market Dollars	Total Market Margin %	Potential ABC Units	Potential ABC Dollars	Potential Gross Margin %	Potential 3 Yr ABC Trend %
1.5	1000	20,000	20%	10%	5,000	100,000	20 %	4,000	80,000	20%	10%
3.0	2000	45,000	10%	15%	15,000	300,000	18 %	6,000	120,000	18%	35%
4.5	4000	85,000	13%	8%	25,000	550,000	23 %	8,000	175,000	30%	27%

Figure 12.6 Manufacturing Strategy: Market Segmentation

actual output for a cell, our overhead may not be absorbed as it has been budgeted and that will leave us under-absorbed, which affects our profit numbers.

"But, it has been that individual focus on 'localized operations' and producing more hours than we need to satisfy the customer demand that has caused us to have the long lead-times that now exist in the factory," said Steve. "We need to concentrate on improving the overall process and quit focusing on the individual operations, if we ever expect to achieve the levels of performance that have been targeted."

As the project team was gathering information on the process, Paula was setting up the database that would house all the data being collected. She devised a simple spreadsheet design with tabs for each of the product groups according to production process. This way no matter what data they needed for analysis, they were very easy to extract. As each team completed their templates, they were turned into a data entry clerk to load into the database. Once the project team had completed the data gathering, they were ready to develop an understanding of the marketplace.

Robert showed the project team two templates (Figures 12.6 and 12.7) and told them to identify who in the organization had access to the information necessary to complete the requested information. Paula felt that she may be able to extract some of the data from the business system, but most of it would have to come from other sources: "I know I can pull and segregate

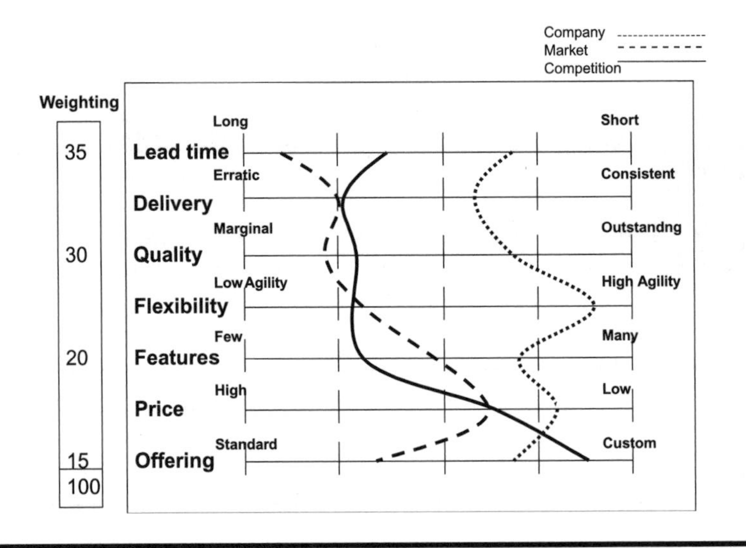

Figure 12.7 Manufacturing Strategy: Competitive Criteria

the sales data based on history, but the total market and potential market have to come from sales and customer service."

Carl stated, "I don't think customer service would be the place to find those projected data. I think product development should have a view of the total market requirements."

"Maybe we need to talk with both," said Robert. "Why don't we bring in Barbara and Samuel, with a few of their product experts, and discuss with them who has access to which data and then work with those experts to complete both of these templates. Remember, we need to have a pretty reliable view of the current marketplace, particularly where specific product opportunities exist, and input from the customers as to where we are competitive and where we are not. This is where much of our design criteria information will be drawn from in order to align with marketing as we get into the Future State Design phase."

As the team completed this final data-gathering effort, they were able to draw a good picture of how Regal stood in relation to the concept of lean. They had an increased understanding of the marketplace through actual data collected from the customers through surveys and interviews. They presented their findings to executive management on August 27. There was not a lot of debate over the numbers (which had been seen in the past), because the process owners Barbara and Samuel had been part of the exercise and had already bought into the validity of the data. Upon receiving approval for the work in phase one, the team was released to move onto phase two — Current State Gap.

13 Current State Gap

The first order of business for the project team was to gain a better understanding of the overall process flow of the factory. They all had their own ideas about how they thought the process worked, but nobody was confident about really knowing for sure; therefore, Robert once again had the project team split into groups. The first group consisted of Paula, Richard, and Steve, who were to create an overall material and information flow map of the operation to gain a better insight into how the physical material and information currently flowed within the plant. They would identify the communication links between suppliers and customers (internal and external), the medium used to present the information, and how often there was an information transaction (Figures 13.1).

The second group was made up of Heather, Carl, and Joseph, who were given the task of generating a Level 0 and Level 1 process map of the current production process. They were shown how to gather the necessary information through a supplier-input-process-output-customer (SIPOC) methodology (Figure 13.2). Robert challenged the teams to gather enough information about the existing process in order to make good decisions in the Future State Design phase, but not so much information that they got bogged down with analysis paralysis. "That is why it is important to stay at a Level 0 and Level 1 for the SIPOC," he explained. "We are trying to describe 'what' is happening in the process, not 'how.' We have targeted two weeks for completion of this effort, per our milestone plan. In order to stay on schedule, we need to be ready to perform root cause analysis by September 10."

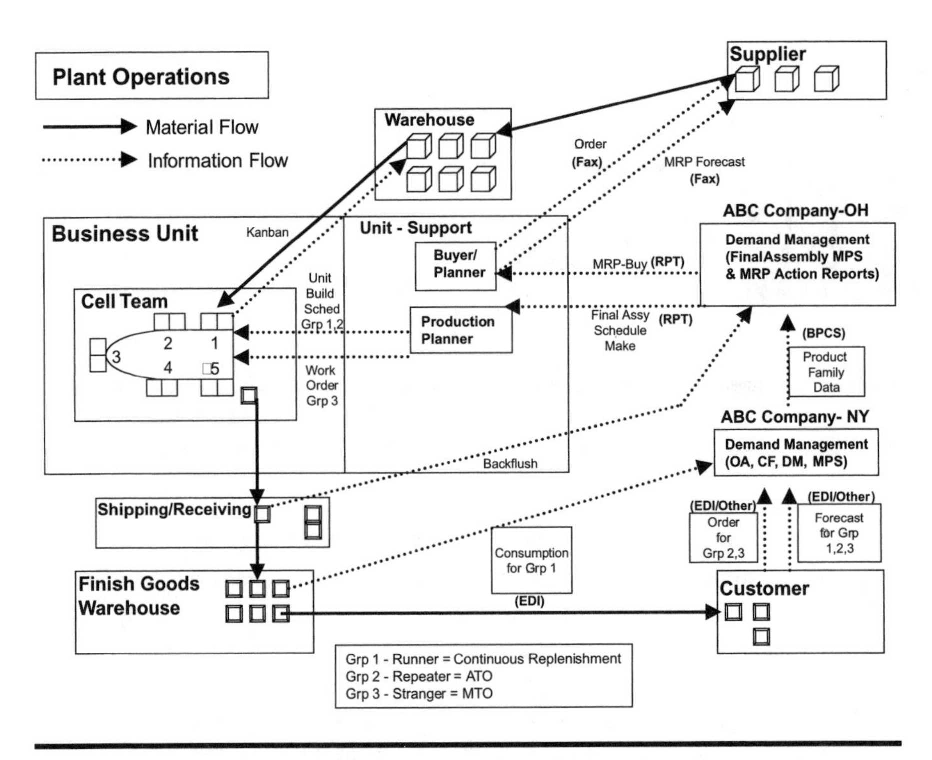

Figure 13.1 Material and Information Flow for Lean Implementation

By 7:30 a.m. Monday morning, both groups were off and running. Paula had taken the lead for developing the material and information flow map. Over the weekend, she had been thinking about how they might approach it. "I think if we identify the production processes that were loaded on the database and review the product families we created during week three, we will have a good indication as to where to start. I think we should lay out the major processes on a white board and represent the primary physical material flows with the color green and show the primary information flows in red."

"Once we have that developed, we can interview those in the process as to the format or medium used to transmit the information. You know … fax, or a hot list, or 3 × 5 card, or electronic, whatever," explained Richard.

"And, once we have these data, we can begin to measure how long it takes for the information to change hands and how often," Steve said.

"Remember," Paula pointed out, "it is extremely important that we verify the data with the process owners or those who work in the process. Maybe we should schedule a meeting next Monday with several of the first-line supervisors — Jake and Ben and possibly Nat — to validate what we find."

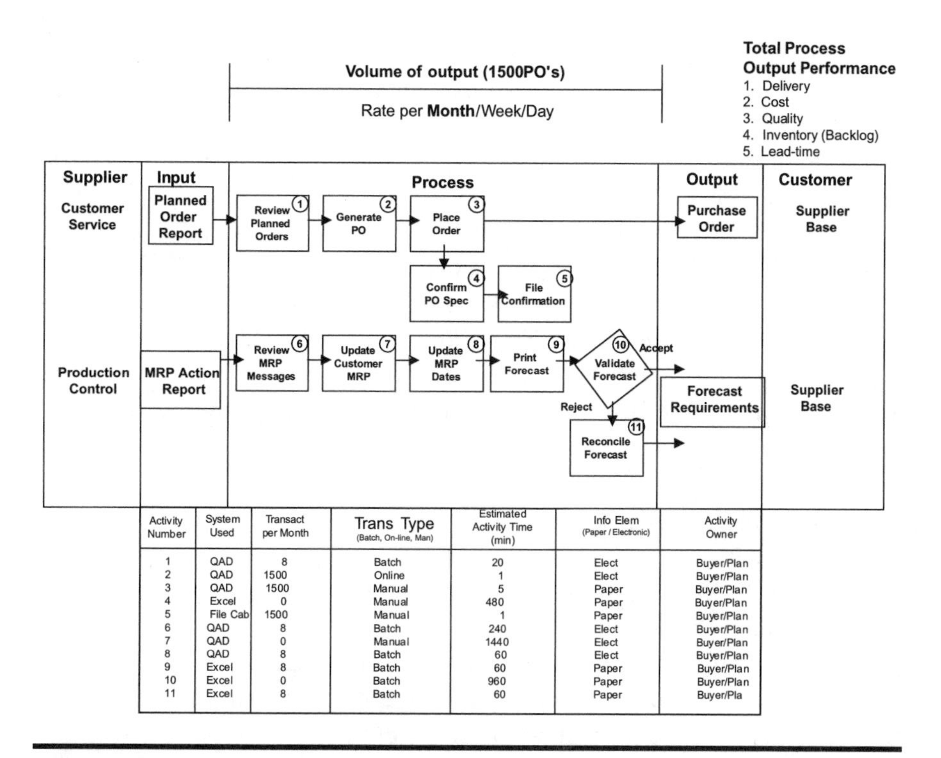

Activity Number	System Used	Transact per Month	Trans Type (Batch, On-line, Man)	Estimated Activity Time (min)	Info Elem (Paper / Electronic)	Activity Owner
1	QAD	8	Batch	20	Elect	Buyer/Plan
2	QAD	1500	Online	1	Elect	Buyer/Plan
3	QAD	1500	Manual	5	Paper	Buyer/Plan
4	Excel	0	Manual	480	Paper	Buyer/Plan
5	File Cab	1500	Manual	1	Paper	Buyer/Plan
6	QAD	8	Batch	240	Elect	Buyer/Plan
7	QAD	0	Manual	1440	Elect	Buyer/Plan
8	QAD	8	Batch	60	Elect	Buyer/Plan
9	Excel	8	Batch	60	Paper	Buyer/Plan
10	Excel	0	Batch	960	Paper	Buyer/Plan
11	Excel	8	Batch	60	Paper	Buyer/Pla

Figure 13.2 Level 1: Supplier Interface

"That sounds like a good idea to me. I will make sure that they will be available early next week to do that," said Richard.

As the first group was making plans for the material and information map, the members of the second group were making similar plans to capture the Level 0 and Level 1 SIPOC data.

"What do you think about starting with outputs by product grouping, identifying the appropriate customers for those outputs, and then documenting the steps in the process that generate those outputs?" asked Carl.

"That sounds reasonable to me," said Joseph. "Then we can list the suppliers for the process and record the inputs."

"I think I would do it the other way around," argued Heather. "I would identify the inputs that trigger the process to begin and then document those suppliers who supply those inputs."

"I can live with doing it that way," said Joseph. "As long as we get done by September 10."

Carl gave Joseph a look. "Once we have Level 0 documented for the overall operations process, we can then break out the level 1 subprocesses into their specific activities," said Carl.

"Keep in mind there may be several Level 1 process flows. We should probably segregate the flows by customer interface, supplier interface, manufacturing, and production/planning control. Also, remember that Robert told us to limit the number of steps to between 6 and 12, so as not to go too deep into the process."

Heather continued, "I think if we approach our largest product family first..."

By the beginning of September, the two groups had made good progress on each type of process map. They documented all the major activities, captured the information linkages, understood how physical material was transported around the shop, and recorded the time required for each process step and the output performance for each product grouping. In addition, they had verified this information with the process owners and received buy-in on the data. When September 10 arrived, it was time to begin analysis of the baseline data.

In order to guide the decision process used in determining (1) the sequence and priority for implementation, (2) which areas were in need of the most help, and (3) justification for additional expenditures, the project team needed to conduct a root cause analysis of the current operating environment. Robert once again had the team break up into two groups. The first group, led by Carl, was to concentrate on the analysis of production and schedule loss. The second group, led by Steve, was to address waste "muda" issues and elements analysis. These two groups were instructed to extract data from their baseline database, process maps, observations, interviews, marketing data, and the lean assessments to generate a clear picture of where wastes could be found in the current operation, the associated causes of the wastes, and their impact on business performance. Each group agreed to a 5-day work window to complete these tasks and expected to finish on September 17, after which they would present their findings to executive management on September 21.

As Carl, Paula, and Richard headed for the war room to begin plotting their next move, Paula asked Carl if he had a clue as to how they were going to come up with this information. Carl replied, "I have been mulling over this one since Robert showed it to us during week two and I think I have a plan. I want to determine a standard output or scheduled amount for each product based on the premise of making today what we need today. I then want to extrapolate the data we review this week on a monthly basis, and then I want to compare the data to the standard. The results are not intended to be additive, but rather show order of magnitude for the problems."

"Did you understand what he just said?" Richard asked Paula.

"I heard him say he had a plan, but after that I haven't a clue!" exclaimed Paula.

"Let me try again," said Carl. "If I have a production area that is required to produce 100 units per day to meet daily customer demand, and the current 'roll-through yield' on that process is 80%, then I have a production loss of 20 units per day, or 400 per month if there are 20 working days in a month. Now, if that same production area has unplanned equipment downtime of 2 hours per day, that would translate to a production loss of 26 units per day, or 520 per month."

"How did you figure that?" asked Richard.

"Well, if we currently run on a one-shift operation of 7.5 hours per shift, that means we need to produce 13.3 units per hour, which I got by dividing 100 by 7.5. Multiply that by the 2-hour loss per day times 20 days per month, and you get 520 units lost per month," explained Carl. "Remember, I did not say the numbers were additive, just that they represented order of magnitude."

"Okay, I guess I understand the production loss, but what about this schedule loss," asked Paula.

"That one took a bit more work, but I think it could work like this," said Carl. "Think about the seven kinds of waste 'muda' that Robert talked about during the lean manufacturing training. He talked about waiting, travel, delays, etc. These kinds of waste can significantly impact an operator who is supposed to be doing value-added work. If I have an operator who is idle 30 minutes waiting for parts or has to spend 20 minutes looking for a fork truck to gather tooling for a setup, that would be a schedule time loss because he is not able to perform value-added work. For example, if I determine that an operator is spending 1.5 hours per day chasing down tooling and his production area needs to produce 100 parts per day, like before, then his potential schedule impact could be 13.3 parts per hour times 1.5 hours per day, which would be a schedule loss of 20 parts per day."

"I see," said Richard. "So, we would gather process performance data about each production area and prioritize the causes based on the magnitude of the impact."

"Exactly," said Carl.

"I hope you two know what you are doing," sighed Paula.

As Carl's group worked their way through the data and began to categorize the causes and magnitude of the wastes, they began to discover some very interesting performance impacts relative to the current operation. It was through the gathering of the data and placing them in this format that they began to develop an appreciation for just how much loss was taking place within the business (Figure 13.3).

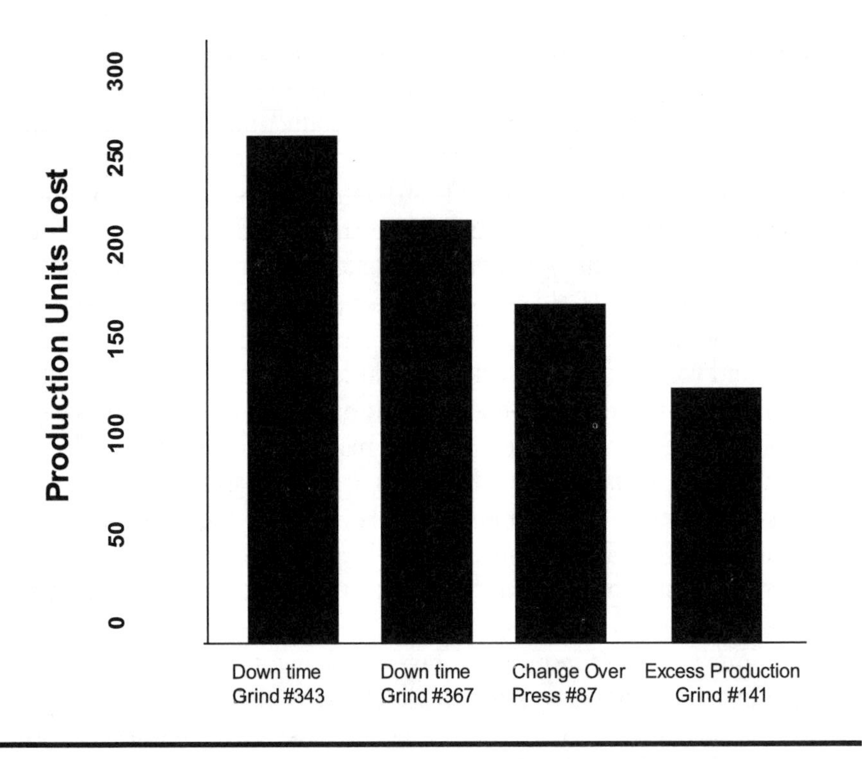

Figure 13.3 Production Loss

Steve's group, on the other hand, did not begin by going to the war room. Instead, they went out to breakfast. Steve offered to treat them to breakfast for a chance to get away to gather their thoughts and collectively decide the best way to approach this next task. "How do you two think we should tackle this next assignment?" inquired Steve.

Between mouthfuls, Joseph said, "I don't care as long as we are done by Friday."

"You are a real stickler when it comes to schedules, aren't you, Joe?" exclaimed Heather.

"What can I say? I'm an accountant," exclaimed Joseph. "I have lived for month-end closings all my life. It's in my blood."

"Are you like this at home?" asked Heather.

"You bet," said Joseph.

"How does your wife stand it?" inquired Heather.

"After awhile, I begin to grow on you," claimed Joseph.

Heather gave up. "What do you think, Steve? How do you think we should proceed?" asked Heather.

Type of Waste	Issue	Element	
Excess Production	Batch Production	SMED	
	Bottlenecks	Service Cell Agreements	
	Curtain Operations	Flex Fence Demand Mgt	
		Metric (Delivery)	
Over Processing	Redundant Systems	DFMA	
	Misunderstood Quality Requirements	FMEA	
	Poor Process Design	Process Capablity	
		Product Flocused Teams	
		Metric (Total Cost)	
Waiting	Down Time	Workable Work	
	Part Shortages	TPM	
	Long Lead Time	Kanban Pull	
		Metric (Lead Time)	
Transportation	Poor Utilization of Space	5S (Housekeeping)	
	Operator Trave Distance	Visual Controls	
	Material Flow Backtracking	Cell Design (Layout)	
		Metrics (Density Ratio/Travel)	
Motion	Low Productivity	Standard Work	
	Multiple Handlng	Graphic Work Instruction	
	Operator Idle Time	Operating Rules/Roles/Resp	
		Cross-training	
		Metric (Productivity)	
Inventory	Long Change Over Time	ABC Material Handling	One Piece Flow
	High Raw Mat'l, WIP, Fin Goods	Material Planning/Control	Work load Balancing
	Excessive Mgt Decisions	Mix Model Manufacturing	Make to Order
		Finish Goods Varaiance	Single Level BOM
		Metric (Inventory Turns)	
Defects	Poor Process Yield	Customer/Suppler Align	SPC
	Employee Turnover	Line Stop Authority	Communication Planning
	Low Employee Involvement	Cell Leader Dev	Cont Improvement Tech
	Limited Process Knowledge	Poke Yoke/Source Inspect	
	Poor Communications	Metric (Quality Yield)	

Figure 13.4 Lean Manufacturing Issue/Element Matrix

"I guess I do not see this as being all that difficult. As I think back to when Robert first showed us the issue/element matrix, it seems to me that it's a matter of identifying each of the current production areas and listing all of the prominent issues in the area. I believe we can gather enough information from the database to generate a substantial list of issues for each area. After having identified the business issue affecting each area, it's a matter of categorizing them according to the wastes that are contributing to those issues (Figure 13.4). From there, we will be able to identify which elements are necessary to fix the business operation problem we are experiencing. How does that sound to you, Joe?"

"Sounds fine to me, as long as we are done by Friday," he said.

Heather and Steve just looked at each other and shook their heads.

By Friday, September 17, each of the groups had been able to create either a matrix or Pareto diagram by production area. They were able to then spend the following Monday and Tuesday morning pulling their executive management debriefing presentation together. The primary purpose

	Quick Hit List			
Production Area	Production Process	Production Machine	Production Loss	Production Impact
D151	Grinding	# 343	Down Time	259 Units
D151	Grinding	# 367	Down Time	235 Units
D180	Punch Press	#87	Change Over	174 Units
D151	Grinding	# 354	Excess Production	141 Units
D134	Final Assy	Line # 3	Operator Idle	134 Units
D134	Final Assy	Line # 2	Operator Travel	128 Units
D156	Machining	# 34	Change Over	121 Units
D156	Machining	# 34	Defective Product	118 Units
D180	Punch Press	# 88	Cycle Time	115 Units

Figure 13.5 Quick-Hit List

of the presentation was to secure management agreement as to the magnitude of the problem, to provide an understanding of the level of potential benefit available, and to explain where the leverage points were to guide the sequence for implementation. In addition, the team had created a quick-hit list of short-term improvements that were discovered during the production and schedule loss analysis.

Robert knew this would be the first real, tangible look by Regal's executive management at how large the gap was and how great the opportunity. It also was the first preliminary view into how much money they may be required to spend to make this program a success. The team recommended that the top five production loss areas be targeted for Kaizen improvements immediately as part of the next phase. The presentation was made jointly by Steve and Heather and was a great success. The project team received approval to advance onward to phase three — Future State Design.

14 Future State Design

On Wednesday, September 22, Brian assembled the team in the war room and congratulated them on a job well done. He pointed out that they had made tremendous progress to this point and were right on schedule. "Now that we have come to an agreement as to where we are, we can now begin the journey of designing where we want to be. This is where the fun starts!"

Robert then explained to the project team, "Our first step will be to create a concept design of the entire factory floor. We will determine how physical material flows between the new manufacturing cells. We will generate a block layout for the plant. We will analyze product demand behaviors and understand the overall resource requirements for staffing and equipment."

"How long will this take?" asked Joseph.

"According to our original milestone plan, we have one week," said Heather.

"After analyzing the part flow between production areas, obtaining a better understanding of process variation, and considering what we now know about the market place expectations, I believe we should expand the target completion to two weeks," stated Robert. "Brian and I have already had this discussion, and he agrees we should extend the delivery date in order to get the job done right. We may be able to make it up in detail design or definitely as part of implementation. This phase is extremely important, because it sets the foundation and direction for the whole rest of the program. Would everyone agree?"

The group as a whole nodded their heads in confirmation.

"Okay, then," said Robert. "I would like to thank Brian for his words of encouragement to the team and in the same breath I would like to ask him to leave so we can get some work done."

Brian nodded his head and made his way to the door.

"Now, what should we do first?" asked Robert. "The specific deliverables for concept design include the number of cells required, an assessment of demand behavior, the new demand management process, plant load profiles, staffing projections, block layouts, product alignment to cells, implementation logic, clarification of design criteria, a weighted decision matrix for layout options, organization chart, business cases for justifying expenditures, and defined exit criteria for each of the cells. Does anyone want to recommend an approach?"

"If it were up to me, I would make sure I had clarification on the design criteria so I knew what we were designing the process to achieve," said Carl. "Then I would want to understand my product demand behavior so I understood which products were high vs. low volume and what kind of demand variation I need to accommodate."

"I agree; that is an excellent starting point," said Robert. "What next, Richard?"

"I would take a shot at aligning which products could be grouped into which cells. I would consider aligning by end customer, high volume, group technology, common routing, material type, etc. I would look at the different options and select the approach that best fits our design criteria," offered Richard.

"I think those are the right items, but I would do them in the reverse order," said Paula. "I think we should agree on the best options first and then allocate products to cells. If we do that, then we can determine the number of cells required, the resource load on each cell, and the staffing needed to support the cell."

"From there we could define our quantitative and qualitative exit criteria for each of the cells for the implementation audit," declared Heather.

"By then we should have enough information to generate the block layout," said Richard.

"From that point we can begin considering the implementation logic, develop any business case justification required, and generate an overall organization concept, as we will have a framework for the factory," said Joseph.

"I like it," said Robert. "Recognize that, although some of these items can be done in parallel, the first few are really dependent items and should be accomplished first. Does anyone have questions? Then lets get started. I want Heather and Carl..."

Demand Behavior	Runner	Repeater	Stranger
Cost of Goods Sold	50 %	15%	35 %
Volume	5000 60%	2000 20%	Low=1500 15%
Per day	1000	250	80
Avg. Order Size	185 units	55 units	5 units
Number of SKU's	Narrow=25	Mid=88	Wide=290
Customer Order Size	Large, Constant	Midsize, Constant	Small
Frequency of Changeover	Low	Medium	High
Order Winning Criteria	Price, Delivery	Price, Delivery	Mfg, Capability

Figure 14.1 Product Demand Behavior

By the end of the first week, the project team had completed all items up to and including the creation of a block layout (Figures 14.1 to 14.4). As they approached the second week, a significant amount of discussion ensued around the implementation sequence and generation of an organization concept.

#	Design Criteria	Criteria Weight	(Option 1) Functional Layout	(Option 2) By Product with Stranger Area	(Option 3) By Product
1	Support Regal, Inc., world-class manufacturing vision	4.8	0.8	4.8	4.8
2	Support runner, repeater, stranger strategy	4.7	2.3	4.7	3.1
3	Facilitate linkages to customer	4.8	0.8	4.0	4.0
4	Support simple materials flow	4.8	0.8	4.8	3.6
5	Increase capacity flexibility	4.3	1.4	4.3	4.3
6	Utilize work-cell approach	4.3	2.2	4.3	4.3
7	Reduce non-value-added space	4.3	0.7	4.3	4.3
8	Provide documented work instructions	4.5	3.8	4.5	4.5

Figure 14.2 Option Selection Matrix

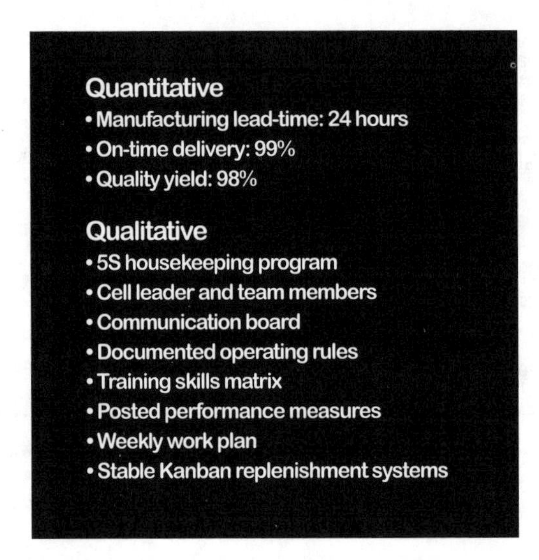

Figure 14.3 Cell: Exit Criteria

"…I don't believe we should start in the shipping area. I know we are having significant throughput loss in the press area due to unplanned down time on equipment," declared Carl. "I know if we start there first, we can continue to gain more short-term benefits."

"I hear what you are saying, but if we selected the customer cell option #2 for three of our assembly cells, then I think that is where we should begin in order to achieve our objective of customer responsiveness," stated Steve.

"Oh, you just don't want to deal with the vendor delivery issues that would arise if we started in the press area first," Carl uttered sarcastically.

"That's not true! We found that our customers for housing and bearing products are most unhappy with our responsiveness. We also found that part of the reason it takes so long is the fact that completed parts sit in packaging and shipping for 3 to 4 days before going out the door. If we can reduce that time to zero by doing the packaging right in the assembly area and sending the product directly to shipping, we could most assuredly meet our customers' expectations of next-day delivery on housing and bearing products," declared Steve.

"Okay, okay, settle down," said Robert. "Let's go back to the reason why we are doing lean in the first place. We have had a customer leave the business due to lack of responsiveness. By losing that volume, we have placed ourselves in a position that will erode operating profit unless we reduce costs, namely

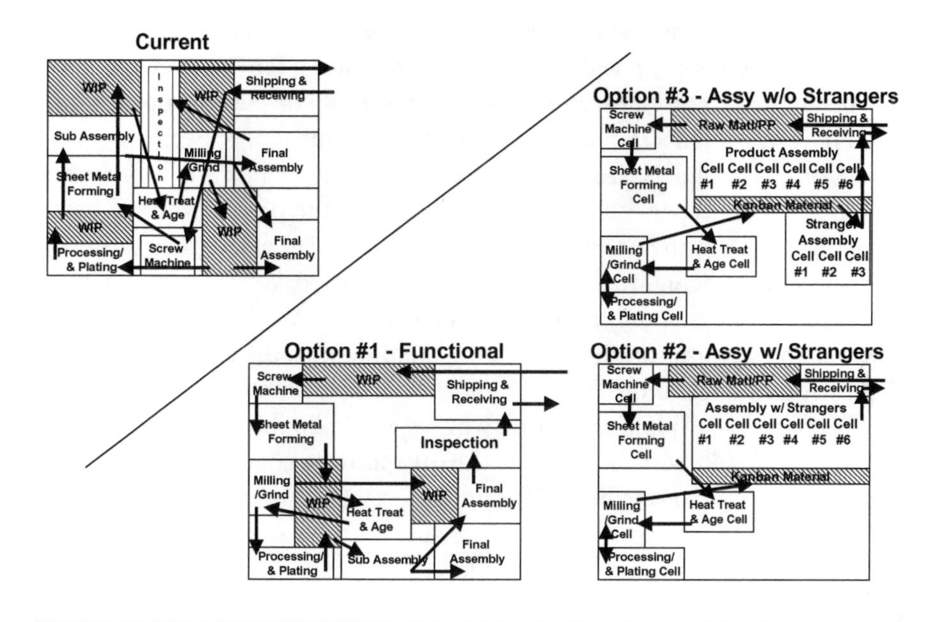

Figure 14.4 Before/After Block Layout

heads. We are not in a survival mode yet, but it is coming. I don't think we want to do anything that will place our supply chain at risk. We have not demonstrated we know how to do lean ourselves; therefore, we don't have much experience to stand on in addressing the current supplier base with lean requirements that we have not demonstrated ourselves. If we can continue to use our supply base as it is currently performing and can minimize risk to the project, I think those are important factors to consider. We are not losing money, cash flow is positive, and we are not asking to spend a lot of money to fund the project at this point."

Robert continued, "What we need is a visible winner and it needs to focus on the external customer. I would agree with Steve. We need to start in assembly and make that area stable. Then we can focus on a fabrication area that has significant production loss and delivers parts to assembly along those specific product families and make that stable. Then we can link the two together using Kanban. Recognize that the fabrication cell may very well make parts for other areas, too, but we can deal with that in detail design and the transition plan. Would this initial sequence make sense to everyone?"

Most everyone nodded their heads in consensus; however, full agreement would have to come later.

"Now, what about this organization concept?" asked Robert. "Is the issue whether we can come up with a proposed organization for this lean operation or whether we want to show a proposed organization for this lean operation?"

"I have had a concern for several weeks now about getting people engaged in the process," said Heather. "I have an issue with publishing a proposed organization concept without having talked with the people who are being asked to change in the process. We have not told them where they fit in. We have not shown them how they are going to be affected. We have not answered what's in it for them if they participate in the process. And, yet, we are creating a proposed organization that may show them doing a different job or show them without a job altogether. I am uncomfortable about doing that."

"Let me try to explain the reason why we would do this and how it should be done," stated Robert. "First of all, the organization concept is to be shared at this stage with no one but this team and executive management. Second, the organization concept is generic in that it portrays what the various roles and responsibilities would be at each level and area within the organization, and the staffing numbers would be an end-state projection based on expected demand levels and the designed staffing to support the demand (Figure 14.5). We need to understand what staffing levels are required to support the business in order to sustain required profit levels. No one will lose their job as a result of the lean manufacturing program. However, if demand falls off and we cannot re-deploy employees to other value-added or improvement initiatives, then a certain number will be laid off."

Robert continued, "We need to let executive management know what staffing level is required to sustain the lean manufacturing environment and, if we are currently staffed heavy, we need to secure more work through increased sales of existing products, new products, or new markets. We do this by arming marketing with a competitive advantage in the market place, so we can grow the business. Remember, this information is obviously sensitive and must be kept under control."

"I understand the need for the organization data, but when are we planning to share it with the people being affected?" asked Heather.

"Good question," Carl said, as he winked at Heather. "We have been going at this for two months now and people are beginning to get nervous. They are asking about what is going on. Why they haven't heard anything, and whether they are going to like this program."

"One of the areas we have not focused on yet is the final plant communication. We have made the opening presentation, we have shown everyone the milestone plan, they have seen the project charter, and they know when we are expected to present our findings. We have been publishing the newsletter

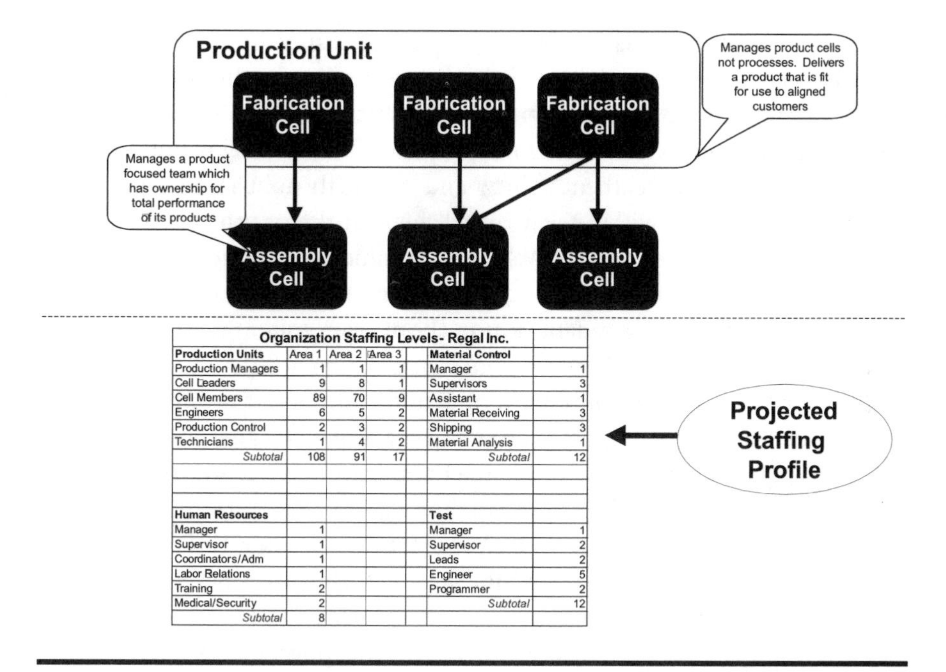

Organization Staffing Levels- Regal Inc.						
Production Units	Area 1	Area 2	Area 3		**Material Control**	
Production Managers	1	1	1		Manager	1
Cell Leaders	9	8	1		Supervisors	3
Cell Members	89	70	9		Assistant	1
Engineers	6	5	2		Material Receiving	3
Production Control	2	3	2		Shipping	3
Technicians	1	4	2		Material Analysis	1
Subtotal	108	91	17		*Subtotal*	12
Human Resources					**Test**	
Manager	1				Manager	1
Supervisor	1				Supervisor	2
Coordinators/Adm	1				Leads	2
Labor Relations	1				Engineer	5
Training	2				Programmer	2
Medical/Security	2				*Subtotal*	12
Subtotal	8					

Figure 14.5 Organization Concept

every other week, and we have been responding to the issue box in order to address individual concerns as we go along," stated Robert.

"What issue box?" asked Joseph.

"The one Brian told them would be put in place and responded to on a regular basis."

"Have we been keeping up with the employee issues box?" asked Robert.

Everyone looked at each other. They had forgotten to assign responsibility for the issue box. Richard ran out to the floor and found the box stuffed full of questions that had not been responded to since day one. He immediately emptied the box and brought the stack of paper into the group. The team was dumbfounded and immediately began cataloging the issues/suggestions and documenting responses to the questions. By about 10:00 that night they had a written response to all the issues and suggestions and had them posted in multiple locations on the shop floor. In addition, they divided up the shop and made plans to visit the shop floor the next day and talk directly with the people and apologize for the project team's mismanagement of the process.

The next morning, when the team visited the shop floor, the majority of them were greeted rather coldly when they inquired about the subject. They were treated to such mutterings as "prima donna," "out of touch," "ivory tower," "not team players." It took the team about three hours to go around and smooth over relations with those employees most potentially affected by the change initiative. After their encounter on the shop floor, Robert gathered the troops and re-opened the discussion about communication and the concept of responsibility, accountability, and authority (RAA). "As we talked about during the first days of the project, when you set up project deliverables and ownership, RAA *must* be established by name for each deliverable in the project. If everybody has responsibility, then nobody has responsibility. I believe we have learned a valuable lesson about clearly stating accountability," said Robert.

"Now, let's talk about this communication plan to be developed," Robert continued. "Contained in that plan is to be a story line that answers some very specific questions: (1) Why are we changing? (2) What are we changing? (3) Where are we now? (4) What's in it for me? At this point in the project, we cannot answer these questions. We are getting closer to being able to answer these questions, but we are not there yet. However, by the end of this phase, we will know these answers and will present them in the plant-wide communication that is targeted at the end of this phase."

"Does it make sense to wait until we are three months into the project before we engage the people with this issue?" inquired Heather.

"I think it is a matter of keeping with each person's role for the project," said Steve. "Think about it. We have brought the process owners in every step of the way as we have gone through each phase of the project. We have gotten their input and buy-in on the validity of the data and not made changes without their concurrence. We have not made any changes to the operational level yet and won't until we begin implementation with the Kaizen events. Our shopfloor operators have not been affected, and when they are, they will be designing their own work areas. We will have done some of the up-front analysis and may have changed what parts are made where, but they will be involved every step of the way when changes are made in their areas on the shop floor."

"I hear what you are saying. I just want to make sure we don't lose sight of the people, because I believe their acceptance or rejection of this project could have a major impact on whether we are successful or not," stated Heather.

All the people in the room nodded their heads in agreement.

"Good, now it's time to begin step two — detail design," said Robert. "The deliverables from this effort will feed directly to the implementation teams for the Kaizen event. For each of the cells, we will be generating a takt time, cell workload, equipment requirements, estimated resources, assigned product mix, SIPOC, cell design guidelines, and potential measures. This effort will save us a tremendous amount of time during implementation. Does anyone have a suggestion as to the best way to get through this? We currently have identified nine assembly, six fabrication, and three service cells for a total of 18 manufacturing cells."

"I would recommend we split the team between assembly and the rest," suggested Carl. "I could lead the assembly team and Richard could take fabrication and service, since we gathered the data from those areas initially."

"That works for me," said Richard.

"That's fine with me, as long as we are finished in two weeks," Paula said, as she nudged Joseph in the arm. The team burst into laughter.

Because there were no objections, the teams were off and running. They jointly created some of the templates, so the information was presented in a uniform manner. They captured all the demand data in order to:

1. Develop a designed daily production rate for the takt time calculation.
2. Generate the daily product-mix schedule required for the cells.

From there, they generated a SIPOC process map for each cell so that all the part numbers for each cell had an identified supplier/customer and any special material handling or processing requirements could be identified. Once they had the required steps in the process documented, the teams captured the current work content for each operation for each part number. This allowed them to calculate takt time, rough out the equipment loads, and to project potential staffing requirements for each of the cells (Figures 14.6 to 14.9).

In addition to conducting a physical flow data analysis for the cells, the teams developed design guidelines for each cell, definitions for the potential measures at the cell level, and a potential organization concept at the cell level. The team spent every bit of the next two weeks designing, calculating, discussing, and debating the design of each of the cells. As the end of the second week drew to a close, the project team was beginning to feel pretty good about what they had developed. An *esprit de corps* was beginning to set in. They were becoming of one mind about the project and generating real excitement about the upcoming implementation.

On Friday, October 15, Robert began shifting the team's focus away from the very detailed, tactical level to a broader, more strategic level. He told them, "We need to spend the last two days developing the transition strategy and implementation plan for the program. The transition strategy should address how we are going to implement the program without shutting down the

Product SKU	Jan	Feb	Mar	Apr	May	Jun	Total	Cum %	Run Cum	Avg Demand	Avg Daily
A	4543	4553	5635	7645	5224	4345	31945	27.37%	27.37%	5324	266
B	3223	2343	4342	3234	4265	3465	20872	17.88%	45.25%	3479	174
C	3567	3675	4675	2565	3564	2766	20812	17.83%	63.08%	3469	173
D	2565	2564	2453	2675	2568	2385	15210	13.03%	76.11%	2535	127
E	2345	2453	2342	2456	2345	1526	13467	11.54%	87.65%	2245	112
F	1253	1646	2174	1526	1412	1454	9465	8.11%	95.76%	1578	79
G	535	353	754	234	567	343	2786	2.39%	98.14%	464	23
H	163	123	123	235	234	153	1031	0.88%	99.03%	172	9
I	76	54	87	100	98	98	513	0.44%	99.47%	86	4
J	100	132	87	56	45	65	485	0.42%	99.88%	81	4
K	34	4	64	23	6	7	138	0.12%	100.00%	23	1
Total	18404	17900	22736	20749	20328	16607	116724				

Figure 14.6 Product Demand Analysis

				Assembly Cell #1						
Product SKU	Jan	Feb	Mar	Apr	May	Jun	Total	Cum %	Avg Demand	Avg Daily
A	4543	4553	5635	7645	5224	4345	31945	27.37%	5324	266
B	3223	2343	4342	3234	4265	3465	20872	17.88%	3479	174
C	3567	3675	4675	2565	3564	2766	20812	17.83%	3469	173
Total	11333	10571	14652	13444	13053	10576	73629	63.08%		614
			Monthly	Peak Demand	14652				Shift Hours	15.0
			Monthly	Avg Demand	12272				Shift Minutes	900
				Delta	2381					
			Variation	Coeficient	19%				Takt Time	1.23
			Daily	Avg Demand	614				(900/730)	
			Daily	Design Demand	730					

Figure 14.7 Designed Takt Time

business. It should answer whether or not we are going to build product ahead of schedule in order to move equipment and buffer customer demand. Are we moving on the weekend? Will we utilize the Kaizen approach? Will we use push-and-pull scheduling methodology in common resource areas for some of the parts or turn over the whole logistics system at one time? How will we locate, identify, count, and track inventory during the relocation? How do we handle our initial excess inventory outside the Kanban system? In addition to the transition strategy, the implementation plan needs to be documented. It needs to identify the pilot cell, which production cells go second and third, and which product groups we are focusing on first, second, and third, etc." (See Figure 14.10.)

As the project team worked feverishly to complete the task by mid-week, Robert was preparing the executive management team for the final debriefing on Friday. He gave them a preview of what was coming and asked if there was anything they could think of that was of concern that the team should look into before the meeting. Every manager said they were quite pleased so far with the planning effort, and they were very anxious to begin the implementation phase after 12 weeks of planning and analysis.

				Processes					
Product SKU	Work Content (in minutes)	Mill Face	Assemble Housing	Test Trigger Assy	Insert Bearing	Grease Fitting	Spin Test	Package	Total
A	Man Time	0.4	1	0.9	0.5	0.3	0.8	0.5	4.4
B	Man Time	0.4	0.9	0.9	0.6	0.6	0.8	0.5	4.7
C	Man Time	0.4	0.8	0.9	0.4	0.3	0.8	0.5	4.1
	Total Minutes	1.2	2.7	2.7	1.5	1.2	2.4	1.5	13.2
Takt Time 1.23		Staffing	4.4/1.23 =	3.5 people				13.2/3 =	4.4 avg

Figure 14.8 Work Content Matrix

Product SKU	Unit Volume	Mill Face	Assemble Housing	Test Trigger Assy	Insert Bearing	Grease Fitting	Spin Test	Package	Total
A	316	126	316	284	158	95	253	158	1390
B	207	83	186	186	124	124	166	104	973
C	207	83	166	186	83	62	166	104	849
	Workload/day	292	668	657	365	281	584	365	3212
	(in minutes)								
	2 shift operation	900							
	equip/process load	32.4%	74.2%	73.0%	40.6%	31.2%	64.9%	40.6%	

Figure 14.9 Volume Matrix

By the time Friday, October 22, rolled around, the project team had their transition strategy identified, they had their implementation plan documented, and they had a plant-wide communication presentation all prepared

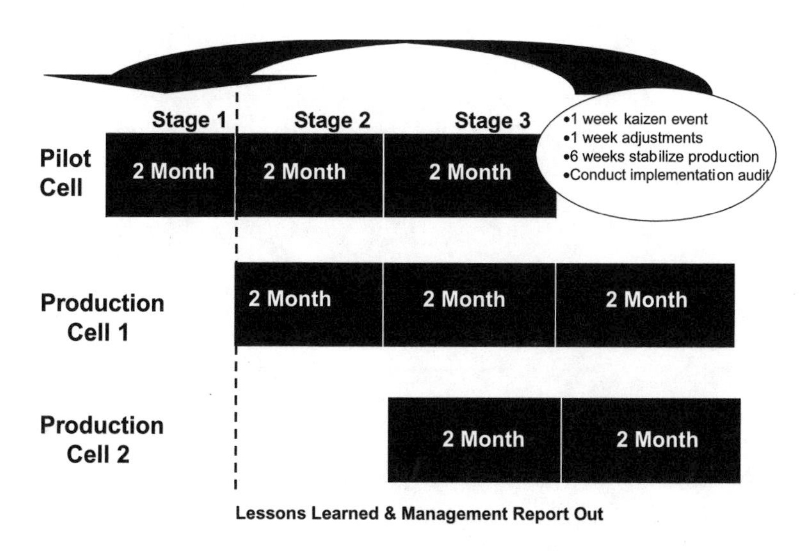

Figure 14.10 Implementation Methodology

for executive management approval. The meeting with management lasted about three hours, including lunch, and at the conclusion of the question-and-answer session, Brian asked that the team be dismissed for an hour while the Regal executive management team contemplated the proposal.

While the team was waiting in the war room, Heather asked, "Why do you suppose they asked us to leave?"

"I don't know," said Steve.

"Maybe they wanted to talk about us behind our backs," said Carl, jokingly.

"I think they just want to make us sweat," said Richard.

"I imagine they just wanted to feel comfortable as a group that this is the right direction and they want to be able to have some free debate among themselves, unencumbered by spectators," said Robert.

After about 45 minutes, they were invited to return to the conference room. Brian welcomed them back and congratulated them on a well-developed and thought-through lean manufacturing proposal for Regal. His next comment was, "So, when can you start?"

15 Deployment

By the time 8:00 a.m. Wednesday morning, October 27, arrived, the team members had already been assembled and were busily making final arrangements for Brian's presentation to the entire employee population of Regal, Inc. It was designed to be a one-hour presentation with a half hour for questions and answers. The team was extremely excited and at the same time nervous about how the lean manufacturing message would be received by the rest of the employees.

Heather was most concerned about how the employees who would be most affected by the process changes would feel about the program. She knew that no one would lose his or her job as a result of the lean manufacturing program. They may be doing different jobs or conducting work in a different manner than they were used to doing, but they would not lose jobs as a result of the continuous improvement efforts. She knew they had already identified a list of existing improvement initiatives, which were not getting done due to a lack of resources. She knew a Kaizen pool was being established for personnel who were released from current production areas so that they could be utilized on future Kaizen events. She realized that Regal sorely needed production engineers and technicians to work on the shop floor and with the supplier base to address lean improvement opportunities. Even though she knew about all the outlets that had been put in place, she was still concerned as to how the people would handle the news. She went up to Carl, lightly touched his forearm, and asked, "Carl, how do you think the people are going to receive the presentation?"

Carl turned to Heather and replied, "I wouldn't be too concerned about it. We have set up several new opportunities if their current positions are being eliminated, and we have assembled a very compelling story as to why the business needs to change. We have shown where they can fit into the new operation. We have demonstrated that when people are working on non-value-added activities they are wasting valuable resource time and energy. We have made it clear that it is the processes they are doing that are non-value-added, and not the individual people themselves who are non-value-added. Unfortunately, I believe what Robert told us is true, that 10% of the employees will embrace the change, 80% will go with the flow, and 10% will fight it all the way. I also think we need to address those fighters as early in the process as possible so they do not ruin it for the balance of the organization. We will need to let them know that we are going in this new direction, and if they don't want to change, we don't think they are going to be happy working here in the future. We need to inform them that we will be glad to help them find a new position someplace else. Hopefully at Blue Iron!"

Heather gave Carl a wide-eyed look. He just gave her a wink and walked back to the war room.

By 10:00 a.m., the stage was set, the podium equipped, the slides loaded, and the crowd gathered for a town hall meeting. The management staff and project team walked in and sat in the front of the room to answer questions from the audience. Brian's presentation was direct, very compelling, and sincerely honest about the current situation and future direction for the company. He talked about the performance gap between Regal today and the benchmark of world-class manufacturers. He described the opportunities for improvement that were identified by the project team. He explained the implementation approach and the timing as to which areas were to be addressed first, second, and third. He showed where Regal stood in the eyes of its customers relative to competitive criteria and by comparison to the competition. He reiterated that no jobs would be lost due to the improvement program, but did explain that many would lose their jobs if there was a reduction in demand for Regal's current and future product base. Brian showed the list of unresourced improvement initiatives that were available for those who were released from their current activities within the operation. He ended the presentation with a thank you to the project team for their efforts over the previous three months and asked for the full support of every employee at Regal during implementation.

Although, most of the audience was quiet at first and did not volunteer any questions, they did not appear to be in shock, either. This was primarily due to the fact that the team had kept many of the key, informal leaders on the company grapevine informed as to what was happening. The team had also kept several influential people involved during the analysis, planning, and design phases in order to validate data and gain concurrence on direction. The questions, which eventually came, were relatively tame and focused primarily on understanding how individual input would be incorporated into their work areas. Brian's response was that as each area was scheduled to come on line, individual inputs would be addressed during that phase of the implementation.

When the project team adjourned to the war room, they were anxious to begin this final phase of the project. They were excited to begin seeing activity relative to all their planning efforts. By the time Robert made it back to the war room, he was pumped.

"Okay," he said. "It is time for us to mobilize ourselves and kick off our pilot cell implementation. Here is how I believe we should proceed. First of all, I want to bring in the selected cell leader and his team of operators after lunch and I want everyone to introduce themselves to the cell team members. Second, I want to congratulate them on being selected for the pilot and let them know that it is a good thing and not a bad thing that they have been selected. Third, I want to brief the cell team on what analysis has been done to this point in their area. Fourth, I want to inform them that they have been scheduled for a Kaizen event beginning next Monday. And, finally, I want to talk through with them the Kaizen event schedule and lean manufacturing principles format."

"Don't you think that that is a lot of information for them to digest in such a short period of time?" asked Richard. "After all, they only received their first introduction to lean manufacturing a couple of hours ago."

"I don't think so," said Paula. "I think these people have been anxious to hear information from us for the last three months and they would be glad to hear as much as we can tell them."

"I agree with Paula," Joseph chimed in. "After all, we had to absorb a lot of information in a short period of time. I think it is time we share the fun with someone else."

The rest of the project team nodded in agreement. They felt the time had come to immerse the rest of this organization in the world of lean manufacturing.

After lunch, the cell team entered the war room somewhat apprehensively. Robert began by introducing himself and asking them to take a seat at the table. The project team members introduced themselves, explained their roles on the project, and congratulated the cell team members for being selected as the pilot cell. After setting the cell team at ease, Steve began the debriefing by explaining the lean assessment results. He then walked them through the overall plant material and information flow map. He then proceeded to describe the production loss Pareto diagram and waste issues/element matrix for their area. He finished by describing the concept design for the plant and showed where their production area fit into the overall layout.

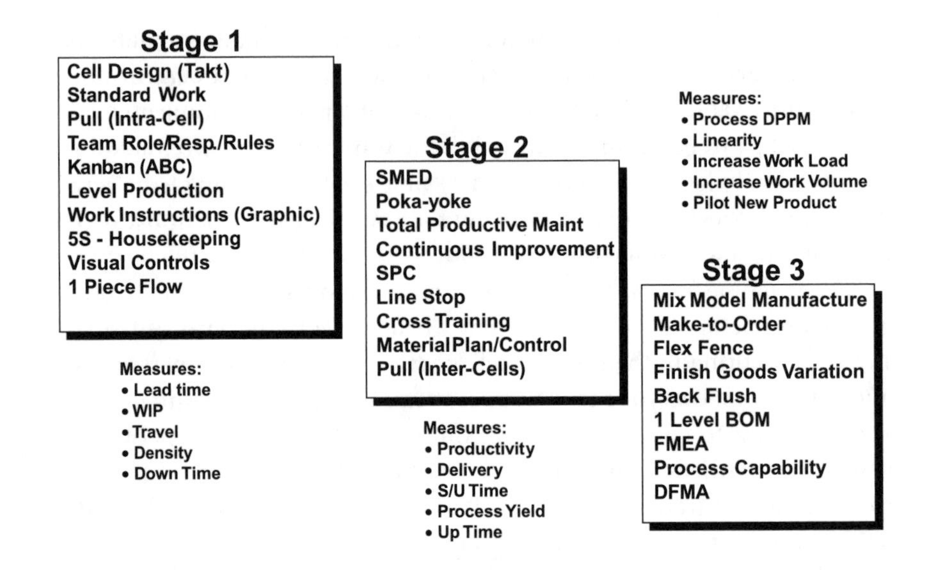

Figure 15.1 Lean Manufacturing Principles

Carl then stood and once again explained the implementation methodology to the team. He explained how there would be a 2-month stage in which the baseline and foundation of the cell would be established. During this period of time, the team would be expected to implement several quantifiable and qualifiable aspects of lean manufacturing within their cell. They would be given time for the process to stabilize, approximately 6 weeks. At the end of those 6 weeks, the cell would be audited. If it passed the audit, approval would be given to advance to stage two of implementation. The same criteria would still apply for both stage two and then stage three. As lessons were learned by the cell team, they would be noted and used for insight during the deployment of subsequent production cells.

"What are these stages you are talking about?" asked Jake, the team's cell leader.

"I am glad you asked that question Jake, because I am about to show you," said Carl, as he hit the button for the next slide.

"The stage one principles are focused on designing a solid foundation for the cell. They are utilized like prerequisites before moving onto those items in stage two. They really set the stage, so to speak." (See Figure 15.1.)

"The first thing we have to do is establish takt time," stated Richard. "Everything starts with takt time, which is nothing more than the designed daily production rate determined for the cell. We are able to determine this from our customer's product demand behavior and the amount of variation we have in that pattern. From there, we need to establish our standard work. We accomplish this through a series of steps: (1) capture the current process flow, identify value- and non-valued-added operations, and time each of the steps in the process; (2) eliminate or reduce the amount of non-valued steps or waste in the process and link the value-added processes together; (3) balance the work load of each operator to the required takt time; and (4) document this as the standard method, sequence, and assignment of work for operators in the cell."

"Once we have an agreed upon a standard work process, we need to identify the component part Kanban requirements and the new rules, roles, and responsibilities for the cell team members," explained Heather. We need to determine the parts and quantities and how they are going to be replenished to the cell. In addition, we will document: (1) the new operating rules for the cell so everyone knows how it is designed to function, (2) the roles each person has as a team member, and (3) the responsibilities each role is accountable to perform."

"Once we have those principles, we can create the graphical work instructions for the cell based on the new standard work," said Carl. "We can then decide what scheduling pattern we want to use to level production through the cell. We look at high- and low-volume demand patterns, setup times, and process quality yields to determine the production level. "

By this time, Robert was beginning to grin from ear to ear.

"With those principles in place, we can now demonstrate the one-piece flow methodology and intra-cell pull concept," said Steve. "One-piece flow means that we no longer build in quantities of days or weeks at a time. We build one piece at a time and pass it onto the next operation without waiting for the rest of the order to complete."

"Finally we establish visual controls and 5S, or housekeeping, criteria for the cell area," said Paula. "The visual controls consist of performance measures on the shop floor, for the shop floor, created by the shop floor, maintained by the shop floor, and owned by the shop floor. Other visual controls include well-marked incoming and outgoing areas, signs describing the cell, and cell boundaries painted on the floor. The 5S concept is a well-organized and structured way to look at housekeeping. Everything has a place and everything is in its place. Everything that belongs is labeled, and the area is cleaned on an everyday basis."

Robert was speechless. His project team had picked up on all the main concepts from stage one, taken them to heart, and was now fully engaged in this new world of lean manufacturing. They had come a long way in the last three months, and he was very proud of the transition they had made. He

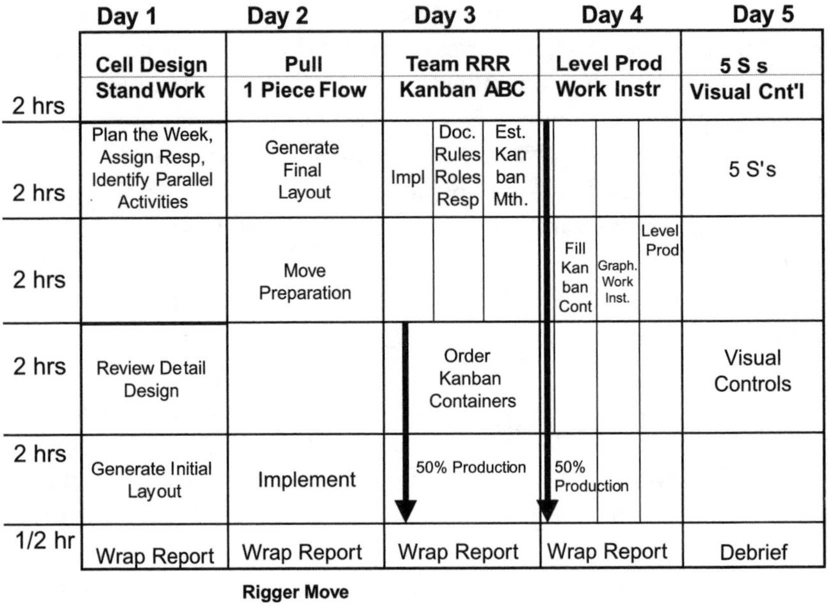

	Day 1	Day 2	Day 3	Day 4	Day 5
2 hrs	Cell Design Stand Work	Pull 1 Piece Flow	Team RRR Kanban ABC	Level Prod Work Instr	5 S s Visual Cnt'l
2 hrs	Plan the Week, Assign Resp, Identify Parallel Activities	Generate Final Layout	Impl / Doc. Rules Roles Resp / Est. Kan ban Mth.		5 S's
2 hrs		Move Preparation		Fill Kan ban Cont / Graph. Work Inst. / Level Prod	
2 hrs	Review Detail Design		Order Kanban Containers		Visual Controls
2 hrs	Generate Initial Layout	Implement	50% Production	50% Production	
1/2 hr	Wrap Report	Wrap Report	Wrap Report	Wrap Report	Debrief
		Rigger Move			

Figure 15.2 Kaizen Event

stood when Paula was finished and began to describe to the cell team the Kaizen event schedule (Figure 15.2). "The Kaizen event schedule you now see before you will begin on Monday. The way it is designed to work is as follows:

"Day One: In the morning, the cell team will receive 2 hours of training on cell design and standard work. This training will prepare your team for the work you will begin on Monday, which will be focused on planning the week, assigning responsibilities, reviewing the detail design analysis, and generating a preliminary layout. At the end of the day, the team will review their status.

"Day Two: The cell team will receive training on material pull and one-piece flow. We are trying to provide the training as you need it and can apply it. In addition, we are trying to keep it concise for ease of retention. On Tuesday, the final layout will be designed, communicated to maintenance, and rearranged. Again, at the end of the day, the team will review their status.

"Day Three: Cell team training continues on Kanban requirements and defining cell team rules, roles, and responsibilities. On Wednesday, the cell team will be doing many activities in parallel. You will be demonstrating the material and operator flow for the new cell. You will be implementing the one-piece flow principle. You will be establishing the Kanban methodology and discipline for the replenishment process. You will be documenting the operating rules, identifying the different cell roles, and describing the responsibilities for each of the new roles. In addition, you will have selected the cell Kanban containers and begun to produce pilot production. Again, at the end of the day, the team will update their status.

"Day Four: The cell team will receive training on level production scheduling and work instructions. You will fill the selected Kanban containers, generate and describe how to maintain the graphic work instructions for the new process, and determine the level production scheduling pattern for their cell. Again, at the end of the day, the team will review their progress.

"Day Five: The cell team will receive final training on 5S and visual controls. The team will use the 5S concept for housekeeping within the new cell, define the criteria for good housekeeping, and establish the manner of audit for housekeeping. In addition, you will determine, design, develop, and deploy three to five critical performance measures for the cell. Two of the recommended measures would be some type of scheduled output adherence and a measure related to process quality. Again, at the end of the day, the team will update their status.

"By the end of the week, the team should have a functioning cell. It won't be perfect and it may not even be exactly the way you really want it; therefore, the following week is available to make changes and adjustments to get it the way you want it. By the end of the second week, we should pretty much have what we want and it is time to let the cell stabilize. There will be agreed-upon performance targets for the cell and an exit criteria established in order to perform an audit. After about 6 weeks, the cell should be performing consistently and have satisfied the exit criteria. It is at this point that we will discuss moving on into stage two. Are there any questions at this time?" asked Robert.

The cell team members looked on with raised eyebrows. They were not quite sure what to make of all this information, let alone how all this was going to be accomplished in five days. Realizing the group was probably in shock, Robert assured them, "I know this is a lot of material, but the project team felt it was important to provide you with a good overview before we just stepped into it next week. Believe me, it will all make more sense as we move into next week."

As the following week progressed, just as Robert said, it all started to make sense. The project team began each day with a snippet version of the training topic for the day. They tracked themselves against the project plan they had developed on Monday. The project team had saved itself a lot of time by doing a thorough job during the detail design phase by determining the demand patterns, takt time, work content times, equipment loads, etc. All this pre-work made getting into the detail layout discussions with the cell teams much easier. The fact that the maintenance resources were on standby to rearrange equipment at the end of the second day and actually got all the equipment moved by the time the team arrived on Wednesday morning blew the cell team away. By the end of the day Thursday, they were actually producing at half rate and demonstrating the one-piece flow concept. By this time, several other Regal employees were becoming interested in what was happening at this new cell area, and by the end of the day Friday, when all the equipment had been painted, floors had been marked and swept, and a communication board was visible, they wanted to know when someone was going to do their area.

As the project team assembled in the war room at the end of the long week, Brian popped his head in and told the team they had done an outstanding job. He had to admit to them, "I wasn't quite sure just how much you really were going to accomplish, but I must confess you have surpassed my wildest expectations. Congratulations! Job well done."

Just as Brain was leaving, Robert arrived to tell the team how proud he was of all they had accomplished. "You, as a team, have come together and learned a great deal from each other. You have acquired knowledge about the current operation, you have applied what you learned about lean manufacturing, and now you are transferring that knowledge to others in the organization. That is where true competitive advantage comes from. It doesn't come from buying a new piece of equipment. It doesn't come from rearranging the furniture. It doesn't come from putting in a Kanban system. It doesn't come from hiring an expert in lean manufacturing. It comes from the strengthening of your organization's ability to respond to customer needs through everyone within the operation.

"Its about equipping everyone with the knowledge of how to be lean and about everyone working to improve the operation a little bit every day. You all have done extremely well up to this point, but just to this point. Implementation is where the rubber meets the road, and you have only just begun. Remember, this is where we start to reap the benefits for all our efforts. Now, go home. You deserve this weekend, but be ready to hit it again on Monday."

16 The Results

As the following weeks and months passed, the lean manufacturing program began to spread across the entire operation. In a matter of 6 months, they were able to bring eight cells up through stage one, and six of those cells also qualified through stage two. None of the cells attempted stage three. It was felt that stage three would be addressed when all the fabrication, assembly, and service cells were functioning at a stage one level. If they kept to the original implementation schedule, all 15 cells would be functioning at stage two within one year of their initial deployment.

Not only were the cells satisfying the exit criteria for qualitative aspects through visible evidence of SMED, TPM, 5S, standard work, Kanban, etc., but they were also affecting the quantitative criteria through greatly improved operational performance. They were approaching between 25 and 35 turns on work in process; on-time delivery output was consistently at 98% every day; and they achieved predictable manufacturing lead-times for products that were measured in hours not weeks. The in-process quality yields were reliably at the 99% level, and labor productivity had increased about 15% across the board without changing any of the current manufacturing processes. These changes in performance had generated tremendous enthusiasm among the employees. Everyone could now visibly see the status of their areas. They knew if they were on plan for the day or if performance was slipping. They were able to identify problems on the communication board as they were happening and make issues visible. The teams were reviewing their operational problems with management on a daily basis for timely corrective action.

It was the culmination of all these individual principles that allowed Regal, Inc., to begin addressing their identified competitive criteria and to align with what their customers valued. They began to exploit their competitive weapons in the marketplace and challenge the competition, which in turn was being reflected by some changes in their bottom line and the securing of additional work that was not planned for the year. Current quarter sales were sharply up over plan, primarily due to the securing of a new order for high-performance pistons from Viscon Company. Regal went head to head with Blue Iron and won a 2-year contract based on their responsive lead-time and reliable quality performance.

These successes would have been short lived and unsustainable if the project team had not realized how important it was to institutionalize the new ways of working, thereby making it extremely difficult for the organization to slip back into the old way of doing business. They knew that by standardizing the work and making the operating processes exactly the same, they would remove a source of variation from the process and enhance output quality. Then, no matter who did the job, by measuring the process performance to a standard for time and output it was very easy to recognize abnormalities as they occurred. This allowed for greater control on the process output and timely feedback for corrective action. Finally, they were astute enough to recognize that rewards for the multiple skills attained by individuals to support the flexible work environment were crucial, as were rewards for consistently achieving and beating performance targets. They offered increased pay for increased skill and rewarded cell teams with performance bonuses on a monthly basis if they consistently achieved target and/or improved the process.

Instituting these changes allowed Regal to become a world-class manufacturing entity over the next couple of years. They had the tenacity to stay the course and ride out the short-term pains for long-term gains. Once they had their shop floor under control, Regal had a working model for suppliers to view so they could begin to deploy similar methodologies within their own factories. Many of Regal's key customer accounts were able to see a significant change within the operation and got a glimpse of where the company was going. This peek into the future impressed these key accounts and helped secure additional new product orders in the future.

For Regal Inc.:

- Consistent *leadership* provided the direction and resources needed.
- *Involvement* of the entire Regal organization allowed them to succeed.
- A lean *road map* helped them stay the course through rocky terrain.
- A passionate *desire* overcame all obstacles along the way.

You, too, can be successful on your path to becoming world class, just like Regal. It is all just a matter of following the right steps.

IV

CASE STUDIES OF LEAN MANUFACTURING PROGRAMS AND PROJECTS

Part IV shows how six different companies deployed lean manufacturing within their facilities (the names and places have been changed to protect competitive confidentiality). Each case addresses a different level or aspect of a lean implementation, but they all follow the same outline in regard to company background, drivers for change, the approach utilized, benefits achieved, and lessons learned. In addition, there are testimonials at the end of each case that provide the reader with some insight into the perceptions of employees experiencing this changeover to a lean environment.

Case Study A: Operations Redesign Program

Company Profile

Located in an industrial park within the city of Juarez, Mexico, resides a manufacturer of uninterrupted power supplies (UPS) for computers. This manufacturer was part of the *maquiiadora* system utilized by many multinational companies as a source of low-cost labor for products. This particular facility was one of many sites owned and operated by a company called Unity Electronics. This primary manufacturing location was contained within a 90,000-square-foot facility, with a total employee population of about 850. Their key manufacturing processes included the automated and manual insertion of printed circuit boards (PCBs) and wave solder operations, as well as manual and automated assembly.

Unity Electronics marketed, designed, manufactured, and delivered UPS systems to the computer and communications industry worldwide. The Unity Electronics operation was divided into several different divisions. The division that owned this particular manufacturing site was Silver Systems Group (SSG). SSG generated approximately $250 million in revenues during 1998 by focusing on three major product segments — standby, line interactive, and online units. The Juarez, Mexico, operation was accountable for producing approximately one half of SSG's revenue. The overall operation was divided among three facilities located in Juarez, Mexico; Horton Mesa, TX; and El Paso, TX.

The original facility was built when the initial company, Faucet, was in the low-volume UPS business. Over time, the need arose to expand into an adjacent building to support a growing demand for higher and higher production volumes. The resulting impact was an ineffective factory flow and insufficient dock space to handle high-volume UPS production. Capacity constraints on equipment limited the ability to satisfy customer delivery requirements and provide on-time shipments. The factory had to operate 24 hours a day, 6 to 7 days a week, to meet customer requirements which did not allow any time for recovery or makeup plans should there be line stoppages. Key customers were also requiring additional capacity and flexibility, which could not be met. To remove some of the constraints, PCB assemblies were outsourced and plans were made to transfer production to other, higher cost facilities within the group. Automated insertion (AI) equipment was running around the clock to keep up with production, which allowed for only minimum scheduled maintenance. Aside from the factory, there was a 30,000-square-foot warehouse facility in Horton Mesa, TX, which handled all inbound and outbound material shipments. In addition, there was a peripheral 13,000-square-foot material staging warehouse in Juarez to handle the overflow of materials due to the ineffective flow through the plant.

Drivers for Change

The operational performance of this manufacturing site had not been satisfactory for several end-item customers over a 3- to 9-month period of time. In November 1997, Unity Electronics was purchased from Faucet and internal management consultants from the new parent company were sent to visit the site to conduct an operations diagnostic on the El Paso, Horton, and Juarez facilities. The result of this diagnostic indicated several issues:

1. Unity desperately needed to get control of its demand management process.
2. The company had a serious delivery performance problem (35% on-time to customer requested ship date).
3. Inventory turns were around 2.8.
4. Supplier management and development were really nonexistent.
5. The limited ownership for product performance was scattered throughout the organization.

6. The planning and control of material and information flow were handled through two different MRP systems. In addition, several informal business rules were used to manage work prioritization on the shop floor.
7. A limited number of shopfloor metrics focused only on quality.
8. Many informal fixes were put in place without institutionalizing the improvements.

In addition, Intermax, a primary OEM worth approximately 50% of Unity's business, had recently come in and rated the quality system of Unity Electronics as very poor — "one of our worst suppliers."

With these identified drivers for change, it is not difficult to see what motivated Unity Electronics to pursue a new way of doing business.

Project Background

Based on the above findings, an initial improvement effort was launched in December 1998 and focused on supply-chain management. By February 1999, it became increasingly obvious that significant synergies could be gained for the business if several ongoing initiatives could be combined under one program. By April 1999, an "operations redesign" program (focusing on lean manufacturing principles) was launched which combined a supply-chain management project, a strategic procurement project, and a plant expansion project into one overall program.

The Unity Electronics Unity Operations Redesign (UOR) program was officially kicked off by selecting a multi-disciplined team to focus on redesigning the value stream for the entire operations process. This team focused on two main tasks: (1) developing an overall conceptual design for the new operation, and (2) generating a project implementation plan that significantly improved the company's ability to satisfy all external customer and internal business expectations. Throughout the project, the project team received significant training in both change management methodologies and lean manufacturing techniques for operations management. In addition to concentrating on the longer term perspective, short-term actions (or quick hits) were identified, and improvements were incorporated as quickly as possible during the concept design phase.

Project Scope and Objective

Unity Electronics' UOR program addressed the entire operations process from customer forecast and demand management through factory floor management and scheduling to supplier management and the distribution of finished goods. It included their global strategic procurement initiative and took advantage of the opportunity to set up a "greenfield" operation within a brand-new facility based on a business unit approach.

The primary performance objectives were intended to affect:

- *Customer requested ship date (CRSD),* the company's performance measured against the date first requested by the customer when an order is placed. This is a measure of the entire process of forecasting, finished goods/service level strategy, and engineering and factory performance.
- *Customer promise date deviation,* the company's performance measured against the first promise given to a customer when an order is placed. The promise date may not equal the CRSD.
- *Manufacturing delivery,* a measure of the ability of the factory to build and ship product on its scheduled date.
- *Manufacturing lead-time,* the length of time from procurement of raw materials to completion of finished goods; also, the minimum length of time from customer order to delivery of requested product.
- *Supplier performance,* a measure of a supplier's ability to satisfy delivery, quality, service, and cost expectations.
- *Inventory levels and turns (raw materials, work in process, and finished goods),* the annual cost of sales (past 3 months annualized) divided by month-end inventory levels.

Project Approach

As was stated earlier, the overall approach to the UOR program actually evolved over time. The project initially began with a focus on the Unity Electronics supply chain, from the customer to manufacturing planning and control on the shop floor to the delivery of finished goods to the customer through warehouse distribution. After a few months of working on the project, it was determined that a greater amount of leverage and subsequent benefit could be achieved through the synergy of several projects, so the entire project was broadened to cover all of operations and placed under one program management structure. This new scope covered everything from order administration and supplier interface to manufacturing management and customer interface.

The overall program was split into several individual projects, which were all interconnected via a common purpose through specifically identified objectives. The projects were segregated by major business process to provide focus for the individual teams and their assigned objectives. Each project had an identified leader with assigned team members. The projects were time phased so that the team members who were assigned to initial projects could be reassigned to later projects. By assigning resources in this manner, Unity was able to achieve cross-functional knowledge transfer through exposure across project teams.

The individual projects included:

1. *Process layout:* Aspects dealing with the physical flow, cell design, and final layout for each of the cells (12) and business units (4).
2. *Material planning and control:* Focus on the design and development of the logistics process for planning and controlling the flow of material through the factory and warehouse system to the customer through Kanban pull.
3. *Organization design:* Organization redesign and training programs that included the cell team, cell leaders, business unit managers, and support operations through a structured process of assessment and selection.
4. *Facilities* (the new plant): Construction of a brand-new manufacturing facility.
5. *Tactical procurement:* Deployment of shared EDI with suppliers through EDI/e-commerce, and reduction of the current supply base by 40%.
6. *Total acquisition cost:* Generation of a global supply strategy and supplier development and selection process.

Each project had its own subset of objectives and assigned deliverables, and each team had to report progress to plan for their project every week. Integration between the project team leaders in regard to what they were designing for the new processes was essential; therefore, communication between teams was a constant activity.

Even though each project was managed independently, they all had to follow the same methodology for design, development, and implementation. This methodology had seven distinctly independent steps:

1. *Baseline* would establish a baseline of current performance for the existing processes. This was accomplished by mapping each of the critical operational processes and gathering key performance data on each of those processes.
2. *Desired state* would establish the desired state for the business. The team did this by reviewing the operations diagnostic that was conducted in December of 1997. They also performed a self-assessment on key business processes to determine where Unity Electronics was performing compared to what was considered best practice. They made site visits to other companies who were noted for operating with lean practices. The expected outcome of this step was for the project team to recognize what was possible and to learn from the techniques of others.
3. *Gap analysis* would recognize the gap between where they were and where they wanted to be. An analysis was performed to understand the gap and identify actions to close it.
4. *Concept design* would provide a high-level concept view of the desired state for Unity Electronics, or a future state vision for what the project team collectively agreed they wanted success to look like. It included deliverables such as block layouts, determining the number of cells, what products are made in the cells, number of business units, etc.
5. *Detailed design* would provide a detailed view of the future state. It described all those elements that make the future state a reality and included deliverables such as cell equipment requirements, equipment loads, Kanban sizes, staffing needs, operating rules, material planning and control process at the cell level, cell team member roles and responsibilities, etc.
6. *Implementation plan* would develop an implementation plan and include the time frame, identified deliverables, assigned ownership, transition strategy, and sequence of events to make the future state a reality.
7. *Execution* actually would deploy the implementation plan.

As each individual project's team analyzed and designed their improvements, they were required to receive approval at each step before moving on to subsequent steps. This ensured control of the program. It kept the steering committee engaged in the project and made sure that they bought into the design solutions before going too far with an unapproved design. It also enhanced integration between the projects because the steering committee was made up of cross-functional managers covering all aspects of the business. Therefore, they were the objective third-party view that looked at the solutions from an outside perspective.

When it came time for implementation, the process owners (those who had to live with the new process after the project was over) were in the driver's seat for deployment. The design team was to still remain assigned to the project until the process owner agreed the new process worked and was doing what it was designed to do.

The one overriding strategy was to prove out the mechanics of the new process in the old facility. When the new operational process for the first business unit was stable, then it would relocate to the new facility, thereby minimizing risk and avoiding a double move of equipment.

Execution of the implementation plan had a few key aspects worth noting:

1. The responsibility for execution was handed over to the individuals who had ownership for the new process after implementation, thereby requiring buy-in to the new design before deployment. This reduced the burden of having to "sell" the new design to those on the shop floor.
2. A pilot cell approach was used, by which the implementation initially concentrated on one manufacturing cell, gathered all the lessons learned from that cell, and then carried those onto the next manufacturing cell. This minimized risk to the project and allowed the project teams to collectively concentrate their energies on one pilot cell during the learning stages of implementation.
3. Business units were deployed one by one in accordance with the manufacturing cells they supported. This allowed: (1) the organization changes to take place based around a specific product family, and (2) ownership for all the operational processes that affected that family to be quickly adopted. This in turn accelerated the arrival of benefits at the bottom line for that given product family.

Project Time Line

Date	*Milestone*
December 1997	Supply chain project launched
February 1998	UOR program detail specification
March 1998	UOR project team mobilized
June 1998	Material planning/control design approved
July 1998	Pilot cell detail design approved
August 1998	Current baseline process completed
September 1998	Pilot business unit design approved
October 1998	Global supply chain strategy approved
November 1998	First cell goes live
December 1998	First cell exit criteria satisfied
January 1999	First business unit goes live
April 1999	New plant comes on line

Techniques Utilized

Workshop Training	*Topics Addressed*
Program and project management	Charter, milestone plan, hazards, issue log, protocol, project organization, project file, risk assessment, detail schedule, deliverables, control mechanisms
Change management	Communication planning, reaction to change, resistors
Lean manufacturing (Five Primary Elements)	One-piece flow, standard work, workable work, percent loading chart, forward plan, cross-training, runner, repeater, stranger, takt time, Kanban, ABC material management, 5S housekeeping, pull scheduling, visual control, roles and responsibilities, operating rules, shopfloor metrics, service cell agreements, mix-model manufacturing, P/Q analysis, project-focused management, continuous improvement, routing analysis
Business process redesign	Baseline performance, gap analysis, future state, concept design, detail design, implementation planning, transition strategy
Process value analysis	Supplier-input-process-output-customer mapping (SIPOC)

Benefits Achieved

Delivery Performance (CRSD)			
	4/98 (Pre-UOR) (%)	*As of 2/99 (%)*	*Target (%)*
Runner products	48	98	99
Repeater products	46	97	97
Stranger products	41	90	90
Manufacturing Lead-Time			
	4/98 (Pre-UOR) (hours)	*As of 2/99 (hours)*	*Target (hours)*
Runner products	21	16	11
Repeater products	30	20	15
Stranger products	50	23	25
	4/98 (Pre-UOR) (%)	*As of 2/99 (%)*	*Target (%)*
Productivity	67	77	84
	4/98 (Pre-UOR) (days)	*As of 2/99 (days)*	*Target (days)*
Inventory	180	106	60

Lessons Learned

- Adhere to and constantly monitor meeting times and project deliverables. If a deliverable is going to be missed, immediately address the issue and develop a recovery game plan. In this particular case, it should be noted that the Mexican culture was not attuned to exact time frames and specific scheduled commitments.
- Do not assume a group understands terms being used; rather, verify that they do understand the terms being used. (Communication! Communication! Communication!) Several terms such as *team* and *Kanban* were new to this culture.
- Drive to detail as early as possible in the project to assure knowledge transfer. If the project team can develop the detail schedule, with the appropriate deliverables, in the correct sequence, they are demonstrating understanding. This pre-planning is critical when it comes time to involve other resources outside the project team (e.g., process owners, specialists) for scheduling meetings, verifying information, and discussing design options.

- Utilize project leader integration meetings to ensure that cross-functional team issues are being addressed and communicated. Depending on the project, this should take place at a minimum of once per week. This aspect is critical when multiple initiatives are being undertaken simultaneously.
- Develop and roll out the communication plan early in the process to avoid excessive rumors and speculation. Employees need to be informed that a new project is underway, why it is being done, and how they are being affected.
- Recognize individual capabilities and limitations when assigning project roles. Do not overestimate the abilities of individuals based on their enthusiasm for the project. Verify that they have been allocated the time for their activities and have the expertise to do the job.
- Make sure project protocols and project files are utilized religiously throughout the project life cycle. The project file is the "bible" for the project. It contains the project status, issues, game plan, and evidence of progress. At the end of the project, it provides a guideline for the next team that has to implement a similar initiative.
- Document project roles and responsibilities early in the project. Make it very clear who has ownership for what at the very beginning of the project. Leave no gray areas or extensive overlap of accountabilities. This will save a lot of headaches later in the project.
- Utilize "Belbin" profiles for insight whenever possible. Meredith Belbin's team role profiles provide valuable insight about the makeup of a team and the probability of success. Take advantage of this insight whenever possible.
- Require full-time team members during the design and analysis phase. Part-time teams will only be able to give part-time results. When a project team has only 20% of its team members' time, it is very difficult to maintain team continuity and focus over the life of the project.
- Enlist process owner buy-in to the new redesigned processes. Process owners should be given responsibility, accountability, and authority (RAA) for implementation whenever possible. They will own the process after the project is complete and therefore must agree with the new design. They must accept ownership for the design; therefore, they should be intimate with its deployment.
- Coordinate rollout of the project with top management approval. Top management has ultimate responsibility for what happens at the plant and therefore should approve major changes to the business process that are under their control.

- Train all employees who will be involved in the project, not just the design team. Process owners need to know how the project is being managed, where they fit in, and the overall direction and philosophy relative to lean management.

Testimonials

"From the first diagnostic to the end of the project, it was the steady pressure, honesty, and professionalism of all the teams that delivered success. The constant feedback really helped keep us on the right path."
—*Vice President, Operations*

"By reorganizing the *entire* Mexico Operations organization into cell manufacturing based business units, we expect to see the following measurable results:

"1. Productivity improvements: reduced direct head-count requirements, extensive training programs and CIP programs.

"2. Increased manufacturing flexibility, the nature of cell manufacturing; we will also be heavily cross-trained at the cell and support team member levels.

"3. Management by objective: virtually every department in the facility has been tasked to develop performance metrics by which to assess their performance, including the business units.

"4. Reduction in the cost of quality: we have implemented progressive inspection throughput the plant, reducing the number of inspectors.

"5. Improved health and safety: the focus on cell ownership along with 5S training will improve the shop organization as well as plant cleanliness.

"6. To move from being one of Intermax's lowest rated suppliers to one of the best in less than one year."

—*Director Plant Operations*

"Taking a significant step forward in the program/project management process accomplished a number of positive initiatives:

"1. Clarified the roles and responsibilities of the management, teams, and participants.

"2. Set forth a standard set of operating rules for all the teams to follow.

"3. Provided a message to all of management that the standard processes will be embraced.

"4. Provided a team structure that affords accountability for its members and leadership."

—*Program Manager*

"Unity made significant improvements in their overall quality and manufacturing process. …The score of 76 on this new survey, as compared to survey scores of 65 in April 1998 and 53 in December 1997, … is one of the best scores in the shortest period of time among Intermax suppliers." —*Intermax Quality System Auditor*

Case Study B: Kaizen Event-Based Lean Program

Company Profile

The headquarters for Winterton Corporation, a $1.5 billion a year producer of industrial products, are located about 10 miles west of the Cleveland downtown city limits. Winterton was primarily segregated into three operating divisions, a centralized sales/distribution operation, and an independent research and development facility. The corporation managed 38 individual manufacturing facilities (27 in the U.S. and 11 internationally). They employed approximately 11,700 people and utilized several independent distributors to supply their various product lines to the marketplace. These lines included products such as ballbearings, industrial application chain, couplings, electrical/mechanical components, seals, conveyor track, gears, motors, and hoists.

The Winterton brand name goes back over 100 years. They have the reputation of providing a quality product that lasts. They had been able to build a strong market presence in North America over the years through: (1) good brand-name equity, and (2) acquisition of other businesses. These two elements allowed them to grow into a sizable organization; however, as time passed, the marketplace changed and foreign competition began to erode a significant share of their market. Since Winterton Corp., as a whole, had been operating in a very mature industry, their introduction of new products to support organic growth had been limited over the years. They supported their primary customer base through a finished-goods distribution warehouse system which was good from a responsiveness standpoint when product was on the shelf, but it required a significant investment in inventory to maintain.

Drivers for Change

Winterton Corp. was not the dominant competitor in many of the markets it served. Although some of the individual sites were performing quite well from a financial perspective (cash flow, profit margins, return on sales, etc.), overall they were experiencing problems from an operational perspective.

Several of the companies were having trouble keeping their full product lines in stock on the warehouse shelves within the distribution centers. Because many of these companies had been unable to link up with OEMs for new product introductions, many of their products were at the end of their product life cycles and competing almost entirely on price (similar to a commodity product). The organization design, manufacturing architecture, and material flow methodologies were struggling to satisfy new expectations for operational performance.

The majority of the operations were managing all their products as "batch and queue" through manufacturing resource planning (MRP II). All products were scheduled with the same planning and control process regardless of their product demand behavior. Capacity planning was not utilized well as a management tool, and production orders were usually launched to the shop floor and capacity constraints reconciled at that time. It was not uncommon to find the master production schedule (MPS) managed via sales dollars rather than by production unit.

Visibility on the shop floor in regard to performance to plan for delivery, quality, inventory turns, cycle time, equipment downtime, productivity, etc. was not clearly evident. There was limited tie-in between shopfloor activities and overall business objectives. It was difficult to see where employees were engaged in the operation of the business, due to the lack of feedback as to how they were performing. In response to the need for the entire Winterton organization to be competitive, a significant change in the way the manufacturing sites were being managed was beginning to evolve.

Project Background

In November 1998, Winterton's parent company merged with a second comparable operation. This second operation had spent several years implementing lean manufacturing across most of its companies. Many of these companies had been following an approach that deployed lean manufacturing strictly via a series of Kaizen events or "blitzes." After several years of conducting Kaizens in manufacturing, it had become evident to these companies that many of the lean techniques used by the shop floor were applicable to administrative environments as well; therefore, they expanded the program to cover the entire business operation. This strategic initiative became known as "lean enterprise."

Following the merger, the new parent company expected each of its divisions and companies to adopt the implementation of lean manufacturing within their facilities. When it came time for Winterton to begin its lean program, they adopted the Kaizen event-based approach. In addition, they supplemented the Kaizen event-based approach with the utilization of a 4-day lean class to enhance knowledge transfer to the workforce. This 4-day lean class was designed for those who already had attended an event in order to reinforce those topics covered during the Kaizen event. This corporate-wide project was launched with an initial pilot deployment at two Winterton companies in April 1999.

Project Scope and Objective

Winterton Corporation initially targeted 23 companies in North America for the adoption of lean manufacturing, with a time frame for implementation of April 1999 through December 1999. The companies were identified, selected, and divided up between several internal management consultants who had previous knowledge of lean manufacturing concepts to support the rollout of the lean enterprise program. The lean Kaizen events and 4-day lean classes were scheduled and attendees invited.

There were two primary objectives for the program. The first was to conduct at least one Kaizen event at each of the 23 sites in order to introduce the organizations to the lean concepts and develop Kaizen event leaders. The second was to expose as many employees as possible to the 4-day lean class before the end of December 1999. The lean class target audience included plant managers, manufacturing managers, buyers, schedulers, production engineers, and first-line supervisors.

The overall intent was to jump start Winterton's move toward lean, to educate as many people as possible about lean tools and techniques, and to demonstrate an improvement in operational performance through Kaizen event projects as soon as possible.

Project Approach

Beginning in April 1999, companies began hosting the Kaizen events at their individual facilities and invited attendees from other Winterton facilities to participate in the events. The events were intended to last 4 days and usually concentrated on four separate projects, typically three processes from the shop floor and one administrative process. Each project had a cross-functional team of 8 to 12 people assigned to it. The strategy was for these events to be utilized as springboards of lean activity within the operations. Once a company had conducted a Kaizen event, they were to continue following up with other events as needed to find waste in the business and continually improve the operation.

Typically an event lasted 3 to 5 days, depending on project scope, objectives, and whether the site had previous experience with Kaizen. The first day consisted totally of training and education. It was a mixture of lecture, exercises, discussions, and simulations. The training addressed multiple lean manufacturing topics (e.g., one-piece flow, Kanban, visual management, measures, etc.). The second and third days were the actual Kaizen event itself, during which the teams: (1) baselined the existing process; (2) designed a new process; (3) demonstrated the new process, including the rearrangement of equipment; and (4) re-baselined the new process. On the fourth day, the teams reviewed their successes and developed a follow-up strategy for any remaining "to do" actions.

Before the event, there was some initial discussion with the company president about the current state of the business and what lean enterprise could do for them. After agreeing on the four projects, project team leaders were assigned. These team leaders, if they had not already run a Kaizen event, were required to attend the event of another site in order to gain some experience in managing a Kaizen event.

In addition to the Kaizen events that were being conducted at each of the sites, the 4-day lean class was being delivered in order to reinforce the learning points from the Kaizen events and to expand the knowledge base of Winterton employees. This lean class was being presented to audiences that had already attended a Kaizen event. Because there was a large number of employees who needed to be exposed to the additional lean material in a short period of time, the lean classes were conducted in parallel with the scheduled rollout of the Kaizen events.

This Kaizen event-based approach to implementing lean management resulted in many of the companies achieving demonstrated performance changes during the one-week event; however, several of the companies were not able to sustain that change. In some instances, companies even reverted back to their original practices and level of performance. It would appear there were several reasons for this outcome:

1. A general lack of preparedness existed before the Kaizen event was launched. Much of the data required in order to begin the analysis phase on the second day of the event were not readily available. The clarification to employees as to management's expectations was limited.
2. A great deal of confusion surrounded the entire week-long event as to what they were doing and why. Little up-front communication about why this lean program was important to the business or how it fit into existing business initiatives had been presented.
3. The follow-up on "to do" activities and coordination of multiple assignments after the event proved to be quite a challenge for many of the companies. Open items lingered for weeks and sometimes months. Key resources within the business were overwhelmed with work (especially maintenance and information systems), and decisions about what to do and where to go next were not very clear to people within the business.

Although many companies struggled, several companies were able to achieve improved performance and successfully sustain it. These companies had several traits in common:

1. *Leadership.* There was a constant driving force that overcame apathy and did not let inertia set into the organization. There was an unwillingness to allow statements such as "We can't do that" or "That won't work here" to stop the effort. The tenacity to see it through and a willingness to try new approaches were constants.
2. *Direction.* An overall plan or vision as to what was next or what success looked like when the program had achieved its mission was verbalized. Knowledge about what the next steps might be and an understanding of which technique to use next along that path were evident.
3. *Common goal/objective.* It was established up front with the project team what they were trying to achieve in a quantifiable manner. There was a concentration of their collective efforts on measurable targets. They compared actual performance against those targets, posted the actual results in order to track performance, and were held them accountable for achieving those targets.

4. *Support.* They provided constant coaching and guidance to the project team in order to help them gain confidence with the tools and techniques. There was continual interaction with the team to help them stay on track, which removed the possibility of inertia setting in and provided encouragement through the tough times in order to keep their spirits up.

The companies that exhibited these traits were able to achieve a change in performance and sustain that change. In all cases, someone who held a leadership position within the operation demonstrated these traits. Be it a president, general manager, or vice president, each of them was a driving force for making the Kaizen event-based approach be successful within their plants.

Project Time Line

4/99	*5/99*	*6/99*	*7/99*	*8/99*	*9/99*	*10/99*	*11/99*	*12/99*
KE = 2 (8)	KE = 4 (16)	KE = 4 (16) LC = 3	KE = 4 (10)	KE = 6 (19) LC = 1	KE = 4 (7) LC = 3	KE = 3 (4) LC = 3	KE = 3 (5) LC = 3	— — LC = 2

Note: KE = Kaizen events; numbers in parentheses indicate the number of projects; LC = number of 4-day lean management classes.

Benefits Achieved

Project Type	*Quantity*	*Result*	*Benefit*
Cell manufacturing	12	Cycle-time reduction	30–95%
		Productivity increase	15–40%
SMED	16	Changeover reduction	35–90%
Manufacturing flow	19	Cycle-time reduction	20–90%
Administrative flow	17	Cycle-time reduction	60–90%
Kanban/material pull	9	Inventory reduction	60–90%
Material flow/stores	4	Cycle-time reduction	60–80%

Techniques Utilized

Lean manufacturing (Kaizen event)	One-piece flow, takt time, percent loading chart, Kanbans, material pull, 5S housekeeping, visual controls, problem boards, shopfloor metrics, process mapping, SMED, TPM, Poka-yoke
Lean manufacturing (4-day)	Jidoka, autonomation, Andon, visual controls, just-in-time, takt time, continuous flow, pull systems, standard work, work element analysis, 5S housekeeping, muda, process mapping, Kanban, Heijunka, Poka-yoke, TPM, OEE, big six losses, job instruction training, cross-training

Lessons Learned

- Proclaim a vision and clarify a level of expectation at the beginning of a project. This is necessary to set the tone, generate a focus for common grounding, and help engage all employees in the change process.
- Capture the current performance of identified processes as a baseline, and measure actual performance results that directly align with and impact the bottom line. These should be tangible measures related to changes in the process (e.g., performance to plan of production schedules, levels of inventory, dollars of scrap per product output, productivity of product output/manhour input).
- Conduct planning and analysis of the business before initiating the Kaizen event. An understanding of current business initiatives, where the company is going, and what level of performance is needed for a product and in what markets is crucial.
- Include activities both before and after the week of the Kaizen event when planning for the event. A lean road map is necessary in order to communicate to everyone in the organization that this is a program to stay and not just a "flavor of the month" initiative. Before the Kaizen event is conducted, all the detailed analysis of demand management,

work content, equipment availability, etc. should be completed. After the event, it is important for the project team to know where to go next and what to expect. This is accomplished by creating a road map that looks beyond just the current week.

- Plan on having the lean management experts remain with the project for an extended period of time during and after the Kaizen event to explain, demonstrate, and verify use of the lean manufacturing tools and techniques. The constant coaching and guidance reinforce learning of the tools and verify that knowledge has actually been transferred.
- Recognize that there is an interrelationship between business processes. Business processes function as a spider web of activities, where a change in one process often affects other processes. When deploying lean manufacturing, it is important to keep this in mind.
- Plan a coordinated effort of all lean initiatives across the plant. Projects should be integrated into one overall game plan in order to achieve synergy between the projects and make sure they are all headed in the same direction.
- Announce to the entire organization what is happening, why the company is doing it, and who is involved with the project. When people see activity happening within the organization and do not understand what it is for or why it is going on, they tend to be suspicious of the project and will not readily engage themselves with the effort.
- Include process owners in the Kaizen event when conducting activity in their area. It is critical that they understand the problems being faced and that they buy in to the solution that has been developed. After all, they have to live with the solution when the event is over.
- Institutionalize or "lock down" the new way of doing business through documentation (standard work) and control (performance measures) to sustain the change. If the new process is not recorded, presented for employee training, and monitored as to variation from the design, then the process has a very high probability of returning to the old ways.

Testimonials

"The four-day class should have been done first. It would have given me a better idea what we were doing and why." —*First-Line Supervisor*

"The Kaizen event was exhilarating! We got more done in two days than we have in the past two years." —*Shopfloor Operator*

"I feel like I have been given a hammer, a screwdriver, and a pair of pliers. I have been shown some tools, but I have no idea how to use them." —*Engineering Manager*

"A great deal of havoc is created over a few days and then nobody remains to help clean up the mess after the event. We have this long list of 'to do' items and nobody to do them." —*Shop Superintendent*

"After seeing the lean class, I have a better understanding of the approach and where several of the techniques fit together, but I am still short on how to use the tools and when." —*Quality Engineer*

"The event was great. Who is going to make sure we keep doing it?" —*Shopfloor Operator*

Case Study C: High-Volume-Focused Factory Project

Company Profile

Within walking distance of the city limits of Orlando is a manufacturer of engineered chain products called TyCor Chain. TyCor Chain is actually the combination of two separate facilities, Tyron and Corbin. Each operation had been building chain since the 1800s, but plant rationalization resulted in consolidating the plants in 1991. The Orlando site had been experiencing a reduction in size and head count since the 1970s and had the space available to relocate equipment.

TyCor was a unionized shop, and workers were members of the United Steel Workers local 829. There were 350 employees on the payroll, of which 175 were direct labor. The facility sprawled over 400,000 square feet and was aligned in a traditional factory layout with assembly departments and fabrication departments. The primary manufacturing processes were turning, grinding, cold forming, heat-treating, and the manual and semi-automated assembly of chain.

Over the years, this operation was at one point able to increase sales to over $50 million per year, with a peak of $55 million coming in 1996; however, several years of market decline, changes in ownership, and a lack of investment had left the company with sales revenues of $38 million and dropping profits. Efforts to convince several different owners that a significant investment in capital was required to turn this operation around were not successful, thus leaving them with limited alternatives to improve the operation.

Drivers for Change

Other than the obvious loss in market share, decreasing revenues, and limited profit numbers, this operation was being hit with heavy foreign competition from such unlikely sources as India, China, Singapore, and Taiwan. The foreign competition was producing a product of comparable quality at a cost significantly less than TyCor's. In addition to the external forces for change, there were internal forces as well. When TyCor's new parent company came to visit, it was made clear that this operation needed to make some significant changes to compete as a viable entity within the group. With these identified issues as drivers for change, it was not difficult to establish a motivation for change within this organization.

Project Background

In 1996, a strategic plan was developed to implement "focused factories" throughout the facility. This plan was well thought out, but it was also expensive due to the required investment in new capital. In early 1997, TyCor Chain invited a group of productivity consultants to work with the management team and shop floor to boost on-time deliveries and increase productivity. The project did not go as planned and had a negative impact on relations between management and the shopfloor union work force. This unresolved conflict was still evident when, in May of 1999, TyCor again initiated a company-wide effort to improve the operation through the implementation of lean manufacturing.

The lean effort initially began as a series of Kaizen events. The program officially kicked off in May 1999, with four individual Kaizen projects (SMED, Kanban, and two product flow cells). Three of these projects were selected because they were the right size to quickly demonstrate a change in performance through the support of shopfloor employees. It was considered critical to begin mending relations with the union and try working together through the Kaizen event as a way to jointly improve the operation. The fourth project, Kanban, was considered an integral part of the development of a "focused factory" concept that was to be deployed later on in the year. During the event, target sheets were created for each of the individual projects, and several of the projects were able to demonstrate improvement during the event.

Over the course of the next few months, TyCor Chain continued to launch additional Kaizen project teams across the factory as "islands of improvement." A tremendous amount of energy was being expended and initiatives were getting started; however, they were having trouble completing all the projects, including the focused factory. Worse yet, results were not showing up on the bottom line. In November, it was determined that a change in course was necessary in order to channel the collective energies of all employees and begin to generate results at the bottom line. So, a course was set for implementing focused factories across the entire facility.

Project Scope and Objective

In December 1999, the Director of Plant Operations conducted a review of the program's overall progress and led a discussion of alternative approaches with the President of TyCor Chain and the company's controller. This meeting was used to clarify executive management expectations relative to the lean program and to obtain support for the new direction.

After a lengthy discussion, they decided upon the following actions:

1. Change the course of the project to concentrate on getting the focused factories up and running, rather than spreading the efforts across the operation through the "island of activity" approach.
2. Once the first focused factory was up and running as a good "working model" that satisfied an identified exit criteria, replicate that model throughout the balance of the factory.
3. Scale back the amount of Kaizen projects currently scheduled and complete those projects that had already been started before undertaking any new Kaizen projects.
4. Specify a project management structure with protocol, steering committee, detail schedules, defined deliverables, assigned ownership for deliverables, etc.
5. Change organizational responsibility for the entire set of manufacturing processes that support a focused factory, and select a focused factory manager with ownership for the product from "cradle to grave."
6. Establish operational measures that would demonstrate bottom-line improvements through improved inventory turns, reduced head count, reduced past-due orders, etc. and would hold the focused factory manager accountable for the performance.

By reviewing annual demand volumes for the entire line of chain products, TyCor was able to determine that the highest volume product demand was for their snowmobile drive-chain product line, which was already being implemented but with limited success. Not only was this a high-volume demand product, but it was also a very standard product with little complexity by way of manufacturing processes. In addition, TyCor claimed 88% of the market for snowmobile chain; therefore, it was determined to make that facility the pilot focused factory.

The primary objectives established for the snowmobile focused factory included:

1. Eliminate all past-due orders (5000 strands at the time of project launch).
2. Improve inventory turns from 5 to 50.
3. Utilize Kanban replenishment for all high-volume components.
4. Assign a focused factory manager.
5. Implement a skill-based pay system to replace the current unionized individual incentive pay system.
6. Utilize one-piece flow methodology (one strand of chain) to enhance quality feedback and speed manufacturing cycle time.

Project Approach

As was stated earlier, in December 1999 the overall approach to the lean initiative was altered. Rather than spread the energy of many people across a large area (remember, this facility is 400,000 square feet), the approach was changed to that of establishing a good working model that could be duplicated throughout the factory.

The initial step in this process was to establish a full-time lean team dedicated to deployment of the lean program. This allowed all those individuals who were trying to prioritize activities with part-time resources to establish a single, primary focus — implementation of lean manufacturing. The next step was to concentrate efforts on securing a win. This was accomplished by designating specific product groups or families (e.g., snowmobile chain) toward which the lean team would channel their efforts at developing focused factories. This is not to say that all other Kaizen efforts were put on hold. TyCor just needed to reduce the quantity of Kaizen events that had been scheduled and reschedule them to some later date. This relieved the burden on the organization infrastructure so that activities could now be completed and the focused factory concept could be deployed. The overall concept for the focused factories required several changes within the operation:

1. The ownership for performance of the focused factory was realigned to the entire process of producing chain. Even though assembly cells were at one end of the building and fabrication of components used in the assembly of chain was located at the other end of the building, responsibility for both was assigned to the focused factory manager of that product grouping.
2. In the focused factories, the assembly cell was established first, as it was closest to the customer. There were specific lean principles required for the assembly cell to function, and specific performance levels were expected. As the cell achieved these "exit criteria" and was considered stable, then the fabrication cell was brought on board. (For the snowmobile factory, these events happened simultaneously due to the fact that it had a narrow product line with high-volume demand and dedicated equipment.)
3. When both cells had satisfied the exit criteria and were performing at a stable level, then they could be linked through a Kanban pull signal for part replenishment. This action eliminated the stock room and generating part demand based on MRPII.
4. The last step was to relocate the entire focused factory to its final configuration, thereby completely linking the entire focused factory from both a physical and a logistical point of view. It was done in this manner primarily because of the expense and risk involved in moving the fabrication equipment before knowing which end of the building was appropriate.
5. The individual incentive system used to compensate the workforce had to be replaced because it was driving the wrong behavior and did not fit with the new focused factory concept. Therefore, TyCor management developed a new skill-based pay system. By labor contract, they could do this because they were designing a new work area with completely different operating rules from the rest of the shop. This new system allowed direct labor employees to make the same rate of pay they made under the old system, but only after they attained a specified level of skill. They established one job classification and called it "factory technician."

For the lean project team, selecting the focused factory manager as soon as possible was key to successfully implementing the new structure. The original launch for the snowmobile product group had been struggling. It was decided that a leader assigned to manage the entire manufacturing process would accelerate the implementation immensely. This proved to be absolutely correct. In addition to the organizational change for the focused factory manager, the team had to address union issues surrounding the new pay scenario. Management explained to the union leadership how the new pay system was to work and then put the new position out for bid with a very positive response.

Once the leader was assigned and the pay system installed, the area was off and running. After operating with the new lean processes for several weeks, the performance of the focused factory and the team building among direct labor cell team members had improved significantly. The operators were concentrating very hard on eliminating (for the first time) their past-due orders and fixing quality issues as they surfaced. In an effort to accommodate cross-training needs, the cell team set up a game plan for integrating cross-training the first two weeks of the month and focusing on production the last two weeks of the month. Working in this manner allowed the cell team time to develop its people and still satisfy customer demand requirements.

The lean team was able to rather quickly reposition the focused factory concept from a struggling implementation to a high-flying success by leveraging three elements: (1) focusing the implementation effort around a product grouping, (2) satisfying the needs of the employee through the pay system and training plan, and (3) assigning accountability for the processes that produced a given product in order to improve performance and achieve bottom-line results.

Project Time Line

Milestone Plan

9/99	*10/99*	*11/99*	*12/99*	*01/00*
Project is launched	Area is cleared and prepared	Assembly is moved and production is stable	Component manufacturing is relocated and production ready	Snowmobile focused factory is functioning as a unit

Techniques Utilized

Workshop Training	*Topics Addressed*
Program and project management	Charter, milestone plan, hazards, issue log, protocol, project organization, project file, risk assessment, detail schedule, deliverables, control mechanisms
Focused factory manager: assessment and selection	Candidate self-evaluation, change receptivity profile, Belbin roles, group interview, gap analysis, development plan
Lean manufacturing (Five Primary Elements)	One-piece flow, standard work, workable work, percent loading chart, forward plan, cross-training, runner, repeater, stranger, takt time, Kanban, ABC material management, 5S housekeeping, pull scheduling, visual control, roles and responsibilities, operating rules, shopfloor metrics, service cell agreements, mix-model manufacturing, P/Q analysis, product-focused management, continuous improvement, routing analysis
Lean manufacturing (Kaizen events)	One-piece flow, takt time, percent loading chart, Kanbans, material pull, 5S housekeeping, visual controls, problem boards, shopfloor metrics, process mapping, SMED, TPM, Poka-yoke
Lean manufacturing (4-day)	Jidoka, autonomation, Andon, visual controls, just-in-time, takt time, continuous flow, pull systems, standard work, work element analysis, 5S housekeeping, muda, process mapping, Kanban, Heijunka, Poka-yoke, TPM, OEE, big six losses, job instruction training, cross-training

Benefits Achieved

Metric	*Baseline*	*Actual (01/00)*	*Target*
Delivery	40%	90%	100%
Lead-time	8 weeks	3 weeks	2 weeks
Inventory turns	5	30	50
Space	7450 ft^2	6800 ft^2	—
Head count	21	17	14

Lessons Learned

- Utilization of a full-time "lean team" is necessary in order to establish priorities and consolidate efforts in the same direction.
- Assigning ownership for process improvement along product groupings removes the functional silo view of problems and assigns accountability for performance improvement to one person. This organizational change has a significant influence on how quickly project objectives are achieved.
- Agreeing on expectations early in the project is necessary so that all parties know what they are trying to achieve and what success looks like when they get there.
- Knowing which lean tools and techniques to use when and how plays an influential role in producing bottom-line result quickly.
- Spreading part-time resources across many initiatives leads to misalignment of priorities and the inability to complete all assigned activities. It places undue burdens on the organization and makes it difficult to complete any activities well.
- Assigning the company controller to the steering committee can have a tremendous benefit when it comes time to develop metrics for a project and when it is time to justify the focused factory concept to others in the organization.
- It is best to assess and select the focused factory manager as early in the process as possible, definitely before implementation.
- Individual incentives produce localized optimization, which does not support the lean manufacturing concept. The removal of an individual incentive-based pay system is a must; however, it needs to be done with the complete knowledge and understanding of the union, particularly in light of how it is going to affect an individual's pay.

- Be sure to develop and follow a game plan when changing the work rules and reward system of employees. Think through how to get from point A to point B. It is not enough to have a good solution to a problem. That solution must be executable.

Testimonials

"I have been impressed with the fundamental culture change that has taken place with the implementation of lean in the focused factory. We have a long-standing history of detailed job descriptions working in an incentive-pay structure. The formulation of the factory technician position, which includes responsibility for all equipment and processes along with producing to customer demand, has allowed us to make this culture change with a high level of acceptance from the work force." —*Factory Manager*

"The way the snowmobile and block chain factory is set up is a good idea. The parts are closer together and you can catch the bad parts faster. Everyone helps each other; that is a plus. The big problem so far has been the heat-treat operation's turnaround time on our parts." —*Technician*

"I like what I see with the new snowmobile cell factory. What a great way to do the right thing in building chain. It will really work. I enjoy working this way. We should have done this a long time ago." —*Technician*

"Lean manufacturing is a good idea. Great things have resulted since the beginning of the focused factory. Training needs to be emphasized more. With the experience that we have in the area, people should receive the best training possible." —*Technician*

"Lean is a great concept with potential in quality and productivity. It brings new challenges to the workforce. Our only problem is our heat-treat turnaround time." —*Technician*

"From the initial concept of creating the focused factory, we knew it was going to be a long journey. There have been many challenges along the way and there will be many more. Overcoming those challenges makes for a stronger team environment and work force. Implementation of lean manufacturing takes dedication. You must eat, breath, and sleep lean. If you don't, you are not trying hard enough. Success is your only option." —*Factory Manager*

Case Study D: Kaizen Event-Based Focused Factory Pilot

Company Profile

Located in the western suburbs of Denver, Bel-Ron is a manufacturer of engineered chain products. Bel-Ron began operations at this facility in 1942, and they employ 301 people, of which 211 are members of the local machinist union 1673. The facility is spread across 370,000 square feet and organized around the production of its two main product lines — conveyor idlers and make-to-order chain. Approximately one quarter of the plant has been dedicated to the manufacture of idler products, with the remaining three quarters laid out in a traditional factory flow for the production of make-to-order chain. The primary manufacturing processes within this facility include punching, grinding, forming, heat-treating, welding, painting, and the manual assembly of chain.

Bel-Ron had been able to increase annual sales revenue for the combined product lines to as high as $63 million back in 1995. They commanded a significant amount of available market share, in part because they produced "everything for anybody." Outside competition had never really been a severe problem; therefore, Bel-Ron had always been a source of positive cash flow for the parent corporation. Because the company was being utilized as a "cash cow," investment in the business for maintaining and upgrading capital equipment had been limited. Efforts to convince several different owners that an investment in capital was required to sustain the operation and help it achieve particular business objectives for the future were limited.

Over the years several niche players began to enter the marketplace and siphon off specific market sectors, but this was not considered a major problem because the company was still generating significant cash flow for the business and still had plenty of market share remaining. It was not until the entire industry as a whole began to fall off that significant problems began to surface.

Drivers for Change

Because Bel-Ron was all things to all people, the operation managed all products in exactly the same way. This, in turn, meant that the majority of their products had very long lead-times compared to marketplace requirements. It also meant that management spent a significant amount of time expediting all products through the facility in order to satisfy specific customer delivery dates (which were being missed). It resulted in high unit costs that were squeezing profitability like never before. In addition, Bel-Ron was beginning to receive less than favorable feedback from customers through supplier "score cards" and even encountered unpleasant customer site visits.

Even though Bel-Ron had stable sales revenues at the time, it was becoming increasingly obvious that that situation was not going to last unless something changed. The conveyer idler operation was facing heavy competition from multiple sources. They were a small player in the marketplace and were trying to compete on price and lead-time. Availability and speed to market were the competitive criteria that customers required, and the company was struggling to consistently satisfy this demand. In order for Bel-Ron to regain market share, they needed to reduce lead-time on standard products to five days or less and maintain on-time delivery reliability of 95% or better.

In addition to these market forces for change, there were internal forces as well. In late 1998, Bel-Ron was acquired by a new parent company. When the executive management of the new parent company came to visit, they found an organization that was operating with 1950s production capability, shopfloor layout, management structure, and organizational culture, as well as a traditional manufacturing philosophy. They saw equipment that was not maintained, manufacturing processes that were laid out by functional department, multiple layers of management reporting, direct labor piece-work incentives (remember, this was a union shop), and undisciplined housekeeping practices.

It was clear that this operation needed to implement a significant change in its business practices to be competitive and satisfy new levels of operational performance. With both market share and internal issues as the primary drivers for change, this organization knew it would have to adopt a different approach to manufacturing in order to become a competitive entity.

Project Background

In the spring of 1999, the new parent company began to roll out a strategic program focused on the deployment of lean manufacturing within the corporation. The program was to be rolled out through a series or "wave" of Kaizen events and 4-day lean classes. The lean classes were to be utilized as reinforcement for the knowledge transfer of lean tools and techniques utilized during the Kaizen event. As the employees became more confident in their use of the tools, they would schedule other Kaizen events and strive for further continuous improvement and waste elimination within the facility.

In April 1999, Bel-Ron hosted their first Kaizen event. Attendees from several other sister facilities were invited to participate and learn how to implement the lean program. Bel-Ron had identified four individual Kaizen projects (order administration, setup, cell manufacturing, and assembly flow) for the event. All of these projects were selected with the idler product line in mind. This product had several operational issues relative to both manufacturing lead-times and excess inventory. It was felt that by coordinating several projects along the same product line synergies would develop between the Kaizen projects, and this would have the greatest impact on the idler bottom-line performance.

Project Scope and Objective

In March 1999, the president of Bel-Ron had asked his management team to select four projects for the upcoming Kaizen event. The team looked at their overall business, analyzed the product demand volumes for each of the product groups, and reviewed the operational performance of the two primary product lines, conveyer idlers and engineered chain.

The team selected the idlers because:

1. They represented $10 million of sales (nearly 15% of all revenues).
2. Production of the idlers was already self-contained, with all the manufacturing process on one side of the building.

3. There was significant opportunity to improve performance with limited risk.
4. The engineered chain product line involved a greater number of part numbers, a significant mix variety, and common resources and equipment, and most of the fabrication equipment was not surface-mounted (meaning that redesign or rearrangement would require digging and pouring new concrete). In addition, most of the utilities were laid in the concrete flooring.

The project scope impacted conveyer idler production from incoming raw material to shipping. The manufacturing processes included forming, welding, assembly, and painting. The project objectives included:

1. Reduce inventory levels from $220k to $180k (20%).
2. Achieve on-time delivery performance of 95% or better.
3. Reduce manufacturing lead-time to 5 days or less.
4. Reduce changeover times by 50%.
5. Improve space utilization.
6. Improve responsiveness through flexibility.

Project Approach

Before the Kaizen event was scheduled, participants from sister companies were invited to attend the week-long Kaizen event. The objectives were two-fold: (1) introduce the participants to the Kaizen approach, and (2) have the participants bring an outsider's perspective to Bel-Ron. The teams were staffed with members that represented multi-functional backgrounds and included shop supervisors, operators, union stewards, engineers, managers, etc. By using cross-functional teams, the statement, "That's the way we have always done things," could more easily be challenged.

The Kaizen event was kicked off on the first day with a full day of training and education. This training consisted of a mixture of lecture, exercises, discussion, and simulations. The content addressed the overall Kaizen process, the schedule for the week, and the introduction to several lean tools and techniques (e.g., one-piece flow, Kanban, SMED, TPM, loading charts, process mapping, Poka-yoke).

The operational processes for producing and delivering the conveyer idler product line were divided among four teams. On the second day, each team had to clarify its scope and objectives, map the current process, and gather baseline operational performance data. Once they had collected all the data and spent time analyzing the current process, they were facilitated through a brainstorming exercise by the team leader in order to develop optional solutions to satisfy the agreed upon objectives. Once the new design was agreed upon, it was turned over to maintenance and skilled craft to begin relocating bins, racks, and equipment.

By the end of the third day, the cell area had had enough equipment rearranged to demonstrate the flow of the new process and recognize significant gains in the area of manufacturing lead-time and inventory reduction. Obviously, the inventory was still there, but they had designed a new manufacturing flow that would not require the excess work in process that was currently available. Finally, the teams generated a 30-day "to do" list to manage the remaining outstanding activities (e.g., ordering weld curtains, bleeding off inventory, running utility lines).

Between May 1999 and March 2000, Bel-Ron scheduled a series of follow-up Kaizen events in such areas as SMED, 5S housekeeping, and Kanban to sustain the gains and keep progress moving forward. These mini-Kaizen projects were used to continually reinforce the principles of lean and demonstrate to the workforce that Bel-Ron was serious about utilizing this approach to improve the business.

Several of the first SMED Kaizen project ideas came from the original 4-day Kaizen event. The 30-day "to do" list provided some initial insight as to what equipment and processes needed to be addressed. Teams for the later SMED Kaizen projects were staffed with a cross-functional representation of employees from across the plant. Most of the teams were staffed with the manufacturing manager, equipment operators, production engineer, tooling engineer, and a manufacturing representative from the engineered chain product line. Over the course of the next 10 months, they conducted seven SMED Kaizens (averaging about one every 6 weeks). They addressed welding fixtures, shears, shaft production, roll cutoffs, angle shear operations, die consolidation, and the development of a die-exchange system. These efforts significantly contributed to the reduction of lead-time and increased flexibility within the idler production area.

As the idler team became more confident in their Kaizen approach, its lean effort became much more of a self-perpetuating situation. Rather than formally scheduling a set number of Kaizens per month or quarter, management let the team decide when, where, and how many they were going to conduct. When the first mini-Kaizen event was initiated, the team was skeptical as to whether this initiative was going to last; however, after the event was finished and they saw how much was accomplished and how management was actually listening and reacting to their recommendations, the group was energized. Between July 1999 and March 2000, the idler operation conducted no less than nine mini-Kaizen events (in addition to the SMED Kaizens). These mini-events focused primarily on:

1. 5S housekeeping (to throw out unnecessary items and identify required items)
2. Freeing up floor space (to improve the flow of material, allow point-of-use delivery for 75% of raw materials, and improve the ability to perform line-of-site management)
3. Cross-training (to increase responsiveness and flexibility within the unit)
4. Limiting and controlling inventory (to establish same-day delivery performance on the top six high-volume products and reduce lead-time through the manufacturing area by removing excess work in process)

These-mini events helped to shape the focus and direct the energies of the idler team toward constant continuous improvement.

The team went beyond just factory rearrangement and flow. In October 1999, the team received approval to address the flow of material and implemented a Kanban replenishment system. They started to implement the system at the point of supply for raw materials. The team analyzed the raw material and components parts flow, determined demand behavior for usage, and identified vendors of the material. In December 1999, the teams selected four vendors with whom to develop operating rules and pilot the implementation of Kanban for 24 parts. They developed a visual Kanban process that utilized a "faxban" as the replenishment signal between the supplier and the customer. The approach worked out so well that by March 2000 they had six different vendors on Kanban for 33 individual parts. In the same month, they were able to turn on a Kanban replenishment system for a specified level of finished goods, thereby allowing them to build to the demand of a finished Kanban signal.

Organizationally, changes were made within the idler product line that were precursors to the establishment of the focused factory concept. Initially, Bel-Ron was deploying lean tools and techniques on the shop floor through Kaizen events. In order to establish ownership for the effort and maintain continuity, it made sense to assign someone to manage the overall project, particularly now that all the manufacturing processes were collected together in a cell. This manager had ownership for the people, reported on performance, and facilitated the lean activities. Bel-Ron used this experience and the associated lessons learned from this pilot focused factory as the model for other focused factories that were designed and implemented throughout other areas of the facility during the balance of calendar year 2000.

Project Time Line

Milestone Plan

4/99	*6/99*	*7/99*	*10/99*	*03/00*
Kaizen event is initiated	First SMED Kaizen is conducted	Mini-Kaizens are launched	Kanban system is approved	Kanban system is functional

Techniques Utilized

Workshop Training	*Topics Addressed*
One-day SMED Kaizen	Process mapping, videotape, internal vs. external setup, one-touch methodology, parallel functions (pit crew)
Lean manufacturing (Kaizen event)	One-piece flow, takt time, percent loading chart, Kanbans, material pull, 5S housekeeping, visual controls, problem boards, shopfloor metrics, process mapping, SMED, TPM, Poka-yoke
Lean manufacturing (4-day)	Jidoka, autonomation, Andon, visual controls, just-in-time, takt time, continuous flow, pull systems, standard work, work element analysis, 5S housekeeping, muda, process mapping, Kanban, Heijunka, Poka-yoke, TPM, OEE, big six losses, job instruction training, cross-training

Benefits Achieved

Metric	*Baseline*	*Actual (03/00)*	*Target*
On-time delivery	85%	95%	95%
Manufacturing lead-time	6–13 days	3–6 days	5 days or less
Inventory level (raw materials)	$220k	$140k	$180k
Setup reduction	88 minutes	20 minutes	44 minutes
Space utilization	49,600 ft^2	48,900 ft^2	48,000 ft^2

Lessons Learned

- Assign ownership for process improvement along an entire product family. This removes the functional silo view of problems and assigns accountability for performance improvement to one person and his or her team. This organizational change will significantly impact how quickly project objectives are achieved.
- Understand the overall business plan and where resources have previously been allocated before launching a lean manufacturing effort. Conflicts in business priorities and confusion about operational focus will arise if this issue is not addressed.
- Minimize exposure, mitigate risk, and obtain results to help reveal cause-and-effect relationships by utilizing pilots and applying the lessons learned.
- Identify the cell leader as early in the process as possible to have time to assign ownership for the new manufacturing process.
- Machine operators are often the process experts and a great source of ideas for setup reduction projects.
- Have a dedicated team (project) leader for the transformation to lean manufacturing. The balance of the project team should be dedicated at least 60% of the time.
- Address the issue of cultural change. Communicating the "need for change" is paramount to achieving support for the new way of doing business.
- Follow through on all planned commitments and demonstrate results (even those that were less than successful). This builds trust throughout the organization and supports the "walk the talk" mentality.

Testimonials

"Kaizen worked out well. We worked together, had good results, and reduced setup time. We need to continue this effort to get more done." *—Shopfloor Operator*

"I was very enthusiastic in the beginning and saw some of my ideas implemented, but then it stopped and I became discouraged. Small quantities are hard to get used to." *—Shopfloor Operator*

"It is necessary to have all parties on the same page (management, supervision, operators, etc.), and it is very important that the decisions of the team can be implemented, without significant management intervention." *—Production Planner*

Case Study E: Assembly Production Unit Project

Company Profile

Just outside the city limits of downtown Houston is a producer of industrial application bearing products called AG Bearing. AG Bearing began operations in 1959 and has operated as a union facility with a total employee head count of 550, of which 425 are members of the local 1160. The facility occupies 500,000 square feet, including administrative offices. The operation has followed a traditional manufacturing layout, with individual departments segregated by production process with supervision assigned accordingly. The primary manufacturing processes have included painting, grinding, turning, heat-treating, and manual and automated assembly of bearings.

Over the course of the last 5 years, AG Bearing had been able to consistently sustain top line revenue in the range of $90 to $100 million per year. The company has been very profitable and successful at delivering on customer needs; however, unit pricing pressures, capacity constraints on some assembly lines, lack of manufacturing flexibility, and excessive inventory levels have made it difficult for AG Bearing to improve the overall performance of its operation.

Drivers for Change

One of the issues surrounding the implementation of change at AG Bearing, was the fact that they were making good profits. Their margins were very good for their industry, their revenue line was stable, and they had been able to satisfy product availability requirements of customers through the utilization of a national distribution warehouse system. In addition, as mentioned earlier, this was a union shop that had not experienced a great deal of change in recent history. Under the existing contract, union membership was able to make good money through individual piece-rate incentive and were not really interested in altering that course. This business scenario presented a very difficult situation in which to initiate a change program. When an organization is not in pain and has been making good money for several years, it is difficult to see a need to change how the business is run.

Although the need for change was not visible financially, it was evident operationally, and for AG Bearing the need for change came from several sources. The initial or primary driver came from the new parent company that now owned AG Bearing. The parent company had begun to launch an improvement initiative across all of its facilities and was expecting all of its companies to participate. The second driver came from the president of AG Bearing who recognized that many of the issues that were not visible at the top line were nonetheless extremely visible at the bottom line (e.g., constant overtime, expedited deliveries, significant management intervention, excess inventory). All these issues were visible at the shop floor and very familiar to those who ran the operations side of the business.

In addition to these internal forces, when representatives of the new parent company's executive management came to visit, they made several comments about this facility utilizing the principles of lean manufacturing to improve the operation. These sources of change were the primary drivers behind why AG Bearing launched a lean manufacturing initiative within their organization.

Project Background

The lean manufacturing effort for AG Bearing officially kicked off in September 1999. The plant manager and his staff identified four individual projects to be improved during a one-week Kaizen event. The projects covered a wide range of topics (Kanban, setup reduction, cellular manufacturing, and product flow). These projects were selected because of the business need to build confidence throughout the organization with the use of the Kaizen process and to quickly produce several successful improvements. Baseline performance target sheets were created for each of the individual projects and demonstrated improvements were recorded on the sheets.

Each of these projects did achieve some level of success and provided a kick-start for the lean program. Over the next several months, more Kaizen events were scheduled for specific topics such as SMED, TPM, etc., and project teams were launched across the factory to focus on these specific projects. By November 1999, it was becoming obvious that after three months of effort on the lean project, not much was changing on the bottom line. Many good things were happening. People were fixing equipment, changeover times on machines were coming down, and the manufacturing areas were looking more organized, but any impact on the bottom line was difficult to demonstrate and this was becoming a source of frustration to all employees involved with the lean initiatives.

In an effort to provide focus and demonstrate a bottom-line impact, it was determined that an effort should be launched directly aimed at the bearing assembly operation. A significant opportunity for improvement resided in assembly, and it was the manufacturing process closest to the customer. Quoted lead-times to customers were in the neighborhood of 3 weeks, and work-in-process inventory levels between assembly and fabrication were $7.5 million in just component parts.

A team was selected and dedicated to this lean project for the purpose of establishing what is called, for all intents and purposes, an assembly production unit. An assembly production unit is an organization design based around cells. Ownership for product performance is assigned to cells for the customers they serve; however, ownership is not "cradle to grave" like that of a focused factory. A focused factory has ownership from raw materials to finished goods. An assembly production unit only has ownership back to a work-in-process stores location for component parts. Unlike departments, which are usually organized for specific processes, a production unit has total responsibility, accountability, and authority (RAA) for the products from work-in-process stores to the final customer. The team's overall mission was to implement as many lean principles in assembly as required to bring about a bottom-line change in operational performance.

Project Scope and Objective

In November 1999, AG Bearing mobilized this full-time lean team to design, develop, train, and implement a lean manufacturing environment in assembly.

This team was staffed with three employees, all of whom concentrated their collective energies on the lean initiative. To give the project some structure and a logical sequence for implementation, and to generate a positive improvement as soon as possible, they identified the Excel Bearing product family as the first assembly area on which to focus their efforts. The Excel Bearing product line had been brought into the AG Bearing facility in July 1999 after the plant closure of a sister plant. The Excel Bearing part numbers, tooling, and quality requirements were all unfamiliar to the AG Bearing employees. The manufacturing process documentation was limited; therefore, the training of new operators was difficult. In addition, Excel Bearing's product had been set up to be assembled in a batch-and-queue mode, not a one-piece flow cell; therefore, it was not surprising that the Excel Bearing assembly lines in the AG Bearing facility were not producing at the level of output required to satisfy customer demand. Demonstrated output was about 3500 bearings per day across the four assembly lines on two shifts. The required output was 5000 bearings per day, and that was not being met even with overtime on Saturday and Sunday. The objective was to develop and deploy a lean manufacturing environment that was able to satisfy a customer demand level of 7000 units per day over a 5-day work week without overtime.

Project Approach

In the middle of November, the lean team received training on change management principles, team mobilization approaches, and project management fundamentals. It was recognized early on that the team members had limited experience with managing projects. It was also apparent to the team that an organization that did not recognize a need for change was not going to be easy to change; therefore, the team opted to follow the structured project management approach outlined in their training.

The team developed an agreed-upon charter and milestone plan outlining their project's scope and objectives. The team limited the project scope to just Excel Bearing assembly, and their objectives were to achieve one-piece flow manufacturing on the bearing assembly lines with improved throughput, productivity, reduced quality problems, and reduced inventory levels. Between the initial launch of the project in November 1999 and February 2000, the lean team had a difficult time getting started and showed signs of significant frustration. A formal review of the project's progress uncovered the following issues:

1. By conducting a Belbin role assessment on the team members, it was learned that none of the team members had a strong implementer role preference, which is crucial for a lean manufacturing project. It was determined that the project leader preferred the role of "specialist," which led to difficulty focusing several of the team members.
2. The project team was experiencing difficulty getting launched in a direction and then staying the course. There was a significant amount of confusion as to who was to do what and what direction the project was supposed to be taking.
3. The project was being controlled in an informal manner, without regular reporting and formal status reviews with the project team.
4. The lean manufacturing project for assembly was launched with a project owner who was not able to drive the project. Regular reporting of progress was not requested, and the owner was not really engaged in monitoring the project's progress.
5. The lean team had been experiencing difficulty in getting the shopfloor operators to consistently work with the lean approach. When the lean team was out on the shop floor, the areas could perform exactly as designed. When the team was absent, performance fell off. It was determined that buy-in and ownership for the new lean manufacturing design and ways of doing business had not really taken place for those who actually owned the process, namely the shopfloor personnel and their supervisors.
6. There was a lack of definition as to who had what role and responsibility for the design, development, and subsequent implementation of the lean project. This was one of the primary reasons a lack of cooperation existed between the lean team and the shopfloor management, who were the targets for the change.
7. It was difficult to tell when a cell was actually implemented. The criteria for success or targeted levels of performance were not clear, not communicated, and not tracked.
8. It was discovered by the lean team that much of the tooling and some of the equipment being utilized by the operators were not capable of producing a good-quality product; therefore, the project was set back a few weeks to identify and correct the suspect tooling.
9. The average number of years of AG Bearing work experience for front-line supervision was 23 years. Many of these front-line managers had never worked outside the existing plant and therefore were not aware of any other ways to conduct business.

After some lengthy discussion between the project team and the steering committee, the following course of action was decided upon:

1. The project scope should be expanded to cover the end-state vision for assembly and the remaining product lines in assembly. The initial implementation effort should remain focused on Excel Bearing, but an end-state concept should be developed for the assembly production unit. In addition, a year 2000 game plan for achieving the concept needed to be generated. This schedule was to have assigned actions with dates and a description of deliverables.
2. A new project owner was to be assigned who had a greater vested interest in the successful outcome for the project and would drive it to completion.
3. The poor-condition tooling would be identified and scheduled for reconditioning in a timely manner.
4. The lean team was to engage the shop foreman (process owner) in the design efforts so that the new process would have buy-in and the handoff during implementation would be seamless. The lean team was to be phased out of the Excel Bearing assembly area when the shop took charge of deployment. Only after all exit criteria had been satisfied could the lean team disengage completely.
5. The lean team was to develop a formal project management protocol for controlling the project by establishing a war room, meeting on a regular basis, conducting project status review meetings, and reporting on performance metrics regularly.
6. To engage the entire employee workforce, particularly the front-line supervisors, an overall lean project announcement was to be delivered to the entire employee population.
7. The lean team was to conduct a formal Kaizen event to officially kick off the deployment of the Excel Bearing assembly lines and physically move to the shop floor during the implementation to show support for the implementation issues.

During the month of March 2000, these changes in course for the project were incorporated and the results were tremendous. The new project leader began enforcing discipline with regard to the new lean processes, and improved levels of performance were being sustained in the Excel Bearing assembly cells. Metrics were reviewed in the war room and updated on a weekly basis. The project team met on a bi-monthly basis to review project status with the steering committee and on a daily basis with the project owner during the implementation of an assembly cell Kaizen event.

In addition to the shorter term initiatives, the lean team developed an end-state vision for assembly and produced a game plan that implemented all product lines in the assembly production unit by the year 2001. These assembly cell designs were based on the same principles as those of the Excel Bearing assembly cells in order to build on the lessons learned and experience gained with the pilot implementation.

Project Time Line

Milestone Plan

2/00	*3/00*	*4/00*	*5/00*
Lean team is mobilized	Year 2000 implementation plan is approved	Excel Bearing cells three and four are stable	Excel Bearing cells one and two are stable

Techniques Utilized

Workshop Training	*Topics Addressed*
Program and project management	Charter, milestone plan, hazards, issue log, protocol, project organization, project file, risk assessment, detail schedule, deliverables, control mechanisms
Change management	Communications planning, reaction to change, resistors
Team mobilization	Belbin roles, conflict management, decision making
Lean manufacturing (Five Primary Elements)	One-piece flow, standard work, workable work, percent loading chart, forward plan, cross-training, runner, repeater, stranger, takt time, Kanban, ABC material management, 5S housekeeping, pull scheduling, visual control, roles and responsibilities, operating rules, shopfloor metrics, service cell agreements, mix-model manufacturing, P/Q analysis, product-focused management, continuous improvement, routing analysis
Lean manufacturing (Kaizen event)	One-piece flow, takt time, percent loading chart, Kanbans, material pull, 5S housekeeping, visual controls, problem boards, shopfloor metrics, process mapping, SMED, TPM, Poka-yoke
Lean manufacturing (4-day)	Jidoka, autonomation, Andon, visual controls, just-in-time, takt time, continuous flow, pull systems, standard work, work element analysis, 5S housekeeping, muda, process mapping, Kanban, Heijunka, Poka-yoke, TPM, OEE, big six losses, job instruction training, cross-training

Benefits Achieved

Metric	*Baseline*	*Actual (04/00)*	*Target*
Parts per manhour	6.4	8.5	12.0
Production output	53%	71%	100%
Daily scheduled hours	66	49	40
Defects per million	6758	2646	700
On-time delivery	55%	88%	95%

Lessons Learned

- Assign true full-time team members, not a roster of team members who still have other responsibilities. This is critical to sustaining a common focus.
- Clarify roles with all project participants so that all parties agree to what they are trying to achieve, who is to do what, and what success looks like when they get there.
- Establish a regular, formal project review process early in the project to control the project and keep it on schedule. When hazards or slips in the schedule arise, they must be escalated according to the protocol and addressed immediately.
- Process owners (those who own the process being changed) must be engaged in the project and commit to the new way of doing business before implementation.
- The project owner must be engaged in the project and has to provide the leadership and drive for the project to be successful.
- The roles and responsibilities of everyone involved in the project must be defined, understood, agreed upon, and documented.
- Clearly define expectation and performance targets at the beginning of the project. It is imperative that all parties involved are of one mind as to what a successful project looks like and how it is to be achieved.
- The utilization of Belbin team roles can provide significant insight into the appropriate structure and potential weaknesses of the team makeup.

Testimonials

"I believe changes will happen. We can get some things done." —*Shop Manager*

"We have to keep one-piece flow for quality." —*Plant Management*

"It is important to have all areas of the plant working toward a common bottom-line goal vs. individual initiatives." —*Engineering Manager*

"We discover problems more quickly. Production scheduling is easier. Quality is better due to the move toward one-piece flow. When we make a mistake, only a few parts are affected and the problem is usually caught right away. One-piece flow also breaks up the monotony of batch work — I used to be an assembler in the batch environment. The Kanban ensures that we have our parts available when needed. At Excel, we had a crib attendant that would deliver our parts. We were always waiting on parts." —*Line Leader*

"The Kanban makes it easy for me to get parts. The quick-change tooling is a good idea. I do not have to look for Allen wrenches anymore. Labeled tooling at the press saves me time — I don't have to search for tooling that is labeled. I want to do a good job every day, but I get frustrated when problems arise. …It is difficult to do four-piece flow on a line even if we have just one problem … it forces us back in a batch mode." —*Operator*

Case Study F: High-Volume and Low-Volume Cell Project

Company Profile

Within a 60-mile drive of Los Angeles is a producer of precision bearing products called Monitor Bearing. Monitor Bearing began operations at this site in 1974. They have functioned as a non-union facility, with approximately 290 employees on the payroll. The operation covers 200,000 square feet and has followed a more traditional factory layout. The key manufacturing processes are cold-forming, screw machining, grinding super-finishing, and automated and manual assembly.

In 1997, Monitor Bearing's sales peaked at $34 million of top-line revenue. The primary markets the company serves are heavy-duty truck, construction, and industrial. The balance of their service products have been handled through a corporate distribution warehouse system. Historically, the company has been successful at delivering to customer needs; however, ever-increasing pricing pressures, capacity constraints, more stringent product availability requirements, and customer change orders have been making it difficult for Monitor Bearing to continue operating in the same manner as they had in the past. Efforts had recently been made to deploy continuous improvement initiatives within the facility, but these met with limited success. Even though some investment in capital had been made and the company achieved some benefit, from a traditional cost-savings perspective, they had yet to realize any benefit from the investment in a lean manufacturing initiative.

Drivers for Change

One of Monitor Bearing's major customers purchased several high-volume bearings for its truck transmission manufacturing operation. For as long as they had been in business together, this customer had placed orders with Monitor Bearing in a very lumpy demand pattern but always with 12- to 16-week firm schedules. Within a relatively short time period, though, the OEM customer switched from a traditional firm fixed schedule to a demand for parts based on a scheduled final assembly sequence. This, in turn, caused schedules to change dramatically from 12 to 16 weeks firm fixed to 8 days firm with changes and the flexibility necessary to accommodate weekly adjustments, which were at times quite significant. Their demand pattern characteristically had large quantities at the beginning of the month and then little at the end of the month. These dramatic changes in demand behavior patterns created havoc on the production floor, not to mention the fact that it was a more costly way in which to conduct business. Soon after the schedule changes, the customer began pressuring the company about its unit costs and product availability. In addition to this external force for change, there was an internal force as well. A new parent company was launching a strategic improvement initiative aimed at reducing costly waste ("muda") throughout all of its facilities and was expecting all of its companies to participate. The combination of these two drivers, one internal and one external, is what drove Monitor Bearing to embark on a new approach to manufacturing.

Project Background

Monitor Bearing had a variety of product-demand volumes for its various product lines. Some of the products demonstrated a very high demand volume (e.g., 3000 units per day), and some a very low demand volume (e.g., 2000 units per year). To develop some momentum for their improvement initiative and to arrest the deterioration of the relationship with one of their primary customers, Monitor Bearing decided their highest volume product line should be the first area attacked. This would give them a chance to channel their energies on one specific product family and deploy the lean techniques rather quickly.

Even though they realized demonstrated benefits from applying lean manufacturing techniques to the high-volume products, they recognized that this focus on high-volume product lines only impacted about 30% of sales; therefore, they needed to investigate other opportunities as well. This meant stepping back and looking at the demand patterns of all their end-item products and segregating them by some common factor (volume, market, customer, material, etc.). In doing so, the company concluded that, for their manufacturing environment, the most appropriate choice would be to sort the products by product size first and then by volume, which was dependent on equipment capabilities. By doing so, they were able to divide their entire end-item assembly area into four major product families: (1) high-volume bearings, (2) medium- to low-volume bearings, (3) larger size bearings, and (4) low-volume service. The high-volume bearings encompassed nine end-item bearings; the medium- to low-volume bearings, 67 end-item bearings; the large size bearings, over 400 various end-item bearings; and the low-volume service, in excess of 500 end items. This segregation of product behaviors allowed the company to design and manage the flow of material through assembly according to the demand behavior its the products.

Project Scope and Objective

Initially, the overall project scope and objectives for the lean improvement initiative at Monitor Bearing were pretty much undefined. The company knew that they had to achieve improved performance and that they had to engage themselves with the parent company's strategic effort to eliminate waste. It was just a matter of understanding the tools and having the organization required to make the change.

Monitor Bearing knew they were experiencing difficulty with one specific customer and that the lean tools and techniques they were learning about would be applicable to any of the high-volume cells. So, in an effort to establish a course and set a direction for the company, Monitor Bearing's general manager formulated a target objective in May for three of the identified high-volume cells: "Inventory turns of 12 are to be achieved by each of these cells by the end of the year [December 1999]. After the end of the year, we will determine a plan of action for the balance of the products."

Project Approach

Monitor Bearing officially kicked off their efforts in June 1999, with four Kaizen event projects (order entry, supplier Kanban, cellular manufacturing, and product flow). These projects were selected because they centered around a need to streamline the flow time from customer order to shipment for high-volume bearing products that were currently in production. Monitor Bearing was facing significant cost reduction and on-time delivery pressures from its primary customer and needed to demonstrate improvement quickly.

During the course of the event, each of these projects did achieve varying levels of success and provided the kick-start necessary for the lean improvement initiative at Monitor Bearing. The high-volume bearing cell was able to achieve a significant improvement in output from between 1200 and 1500 units per day to the 2000 per day that were required. They achieved this by analyzing the flow of material, understanding the work content, and balancing the work between stations. In addition, by implementing small batch flow and Kanban pull, they were able to not only improve inventory turns from 5.3 to 12.3 but also arrest a nagging quality problem that was causing them to lose around $5000 per month in the form of scrapped parts.

As the following months passed, more Kaizen events were scheduled and improvement teams launched across the factory. A tremendous amount of activity ensued around plant-wide visual communication of concepts, team accomplishments, 5S housekeeping, equipment clean-up, and implementation of Kanban replenishment for many of the purchased and manufactured parts in both assembly and fabrication. Some plant-wide efforts were initiated relative to single-minute exchange of dies (SMED), which focuses on reducing changeover time, and total productive maintenance (TPM), which focuses on the reduction of unplanned downtime on equipment.

By September 1999, it was becoming evident that the next areas of improvement within the facility were going to be more complex and that the Kaizen projects approach of "islands of activity" used so far was not going to address some of the more substantial business issues necessary for success. So, Monitor Bearing announced the deployment of a full-time lean team to focus on the design, development, and deployment of an overall lean environment for the company. This team was staffed with half a dozen employees who concentrated their collective energies on lean manufacturing activities.

As the lean team became more proficient with the lean tools and techniques, it was time to begin making plans to address the next areas of opportunity within the business, namely the lower volume and higher mix product families.

In January 2000, the lean steering committee held a formal review to assess the progress on the lean improvement initiatives to date and to plot a course for the next year's activity. The organization had made great strides in the area of inventory reduction, inventory turns, and scrap reduction with their high-volume cells during the previous year, and they had increased their output capability by 25%. The amount of customer orders running behind schedule had been reduced to virtually nothing. However, among all these significant achievements, it was recognized that there were still a few outstanding issues that should be addressed by the lean team before tackling the lower volume product lines. After lengthy discussion between the lean team and the steering committee, the following course of action was agreed upon:

1. Lock down, button up, and institutionalize the changes made to this point. Several key changes had been made to the operation, but they had been neither well documented nor completely understood by all the people involved with the change.
2. Define what a completed cell looks like. There were varying opinions as to when a cell implementation was complete, thereby leaving some to feel it was time to move on while others felt there was more to do. This definition of a cell would lead to the establishment of an "exit criteria" for the cell, or quantitative and qualitative elements necessary for the cell to be implemented.
3. Establish an implementation approach that would deploy the identified principles of a cell in stages, the concept here being to incorporate the foundation principles of the cell in stage one and then come back at a later date to implement the principles necessary to bring about a new level of operational performance in stage two.
4. Assign a factory manager to manage the high-volume products of the cell. Supervisors were assigned across departments, which made it difficult to define ownership for product performance and thereby generate continuous improvement.
5. Determine a time line for deploying the above-mentioned actions and develop a game plan for designing and implementing the lower volume/higher mix cells.

In February 2000, the lean team concentrated on documenting the new processes in the high-volume cells, establishing a common definition for a successfully implemented cell, and determining the exit criteria required for a cell to be considered stable. The steering committee took on the action item to work with management on selecting the appropriate people to be cell leaders for the new lean environment. Once these identified principles were implemented and in place for about 4 to 5 weeks, the cells began to exhibit new levels of performance, which paved the way for planning the low- to medium-volume cell. The team addressed this cell differently than the high-volume cells, which had:

1. A limited quantity of end-item part numbers to deal with
2. The same manufacturing processes involved with each product
3. A very consistent operational time from part to part at each station
4. A relatively consistent customer-demand pattern from month to month
5. A limited quantity of high-volume components to Kanban

In contrast, the low- to medium-volume products had:

1. 67 different end-item part numbers
2. Different manufacturing processes and equipment, depending upon the end-item configuration and part size
3. Operational times that were relatively consistent from part to part at each station, but required setup times at each station ranging anywhere from 30 minutes to 4 hours for changeover between product lines
4. A product-demand behavior that varied from 1200 per day to 5 per day, with an order frequency pattern from every week to once per quarter
5. A wide range of component parts with varying quantities depending on the end-item mix, with some of the component parts being used in multiple end items
6. A significantly greater number of machines and assembly complexity

The lean team followed a structured methodology for cell design which captured detailed data about the existing low- to medium-volume product family. Using this methodology allowed them to:

1. Calculate demand quantities per day in order to establish runner, repeater, and stranger product behaviors for low- to medium-volume products.
2. Establish material and work flow patterns by mapping the process and identifying volume percentages between stations.

3. Verify if any of the existing product routings had backtracking or a reverse flow of material.
4. Capture work content times to understand variations between products and between work stations that were designated for the cell.
5. Generate work loads on equipment to see what and how many machines were needed for the cell.
6. Determine takt time for each of the products and in total for the cell. By reviewing the demand pattern, they could calculate a designed daily production rate to accommodate variation for runner and repeater products.
7. Understand how much of an impact existing setup times would have on scheduling the mix of products.
8. Design an appropriate hard-signal Kanban replenishment system to allow for the right raw materials/components being available in the right quantities at the right location.

The final design produced the following results:

1. Equipment was comprised of 20 grinders and three assembly methods.
2. Roles and responsibilities over the entire operation (from work-in-process stores to grinding, boring, final assembly, packaging, and shipment) were defined and clarified.
3. Runner products were dedicated to a particular set of equipment and built to a daily rate.
4. Repeater products were shared across common equipment and built on demand to a replenishment Kanban from shipping. The priority for Kanban orders was first-in/first-out (FIFO). Capacity was allocated based in the designed daily production rate.
5. New equipment was assigned to provide for 20% growth in this segment to accommodate for service and lead-time reduction objectives.
6. Stranger product orders were bundled over a 2-week period of time and scheduled to run twice a month across all available equipment in the cell. Because these products were not sold based on lead-time or unit price, they did not require the immediate turnaround of the runner and repeater products.

Project Time Line

Milestone Plan

6/99	*7/99*	*10/99*	*11/99*	*01/00*	*04/00*
High-volume cell Kaizen event is initiated	Completion of 30-day to-do list	Lean team is launched	Third high-volume cell is operational	Low- to medium-volume project is launched	Low- to medium-volume cell design is complete

Techniques Utilized

Workshop Training	*Topics Addressed*
Lean manufacturing (Five Primary Elements)	One-piece flow, standard work, workable work, percent loading chart, forward plan, cross-training, runner, repeater, stranger, takt time, Kanban, ABC material management, 5S housekeeping, pull scheduling, visual control, roles and responsibilities, operating rules, shopfloor metrics, service cell agreements, mix-model manufacturing, P/Q analysis, product-focused management, continuous improvement, routing analysis
Lean manufacturing (Kaizen event)	One-piece flow, takt time, percent loading chart, Kanbans, material pull, 5S housekeeping, visual controls, problem boards, shopfloor metrics, process mapping, SMED, TPM, Poka-yoke
Lean manufacturing (4-day)	Jidoka, autonomation, Andon, visual controls, just-in-time, takt time, continuous flow, pull systems, standard work, work element analysis, 5S housekeeping, muda, process mapping, Kanbans, Heijunka, Poka-yoke, TPM, OEE, big six losses, job instruction training, cross-training
Cell design	P/Q analysis, process mapping, routing analysis, takt calculation, workload balancing, Kanban sizing, standard work, one-piece flow

Benefits Achieved

Metric	*Baseline*	*Actual (12/00)*	*Target*
High-Volume #1			
Inventory dollars	$484,000	$248,000	$400,000
Inventory turns	5.5	12.3	12
On-time delivery	40%	80%	90%
Scrap percent	3.7%	1.5%	2.0%
High-Volume #2			
Inventory dollars	$407,000	$200,000	$350,000
Inventory turns	9.5	20	12
On-time delivery	33%	88%	90%
Scrap percent	3.7%	1.2%	2.0%
High-Volume #3			
Inventory dollars	$335,000	$1000	$200,000
Inventory turns	4.2	8.5	12
On-time delivery	33%	65%	90%
Scrap percent	0.8%	0.7%	1.0%

Lessons Learned

- Establish a full-time project team to dedicate the resources necessary to focus on and provide support for the integration requirements necessary with an initiative that is managing multiple aspects of lean.
- Clarify expectations early in the project so that all parties know what they are trying to achieve and what success looks like when they get there.
- Recognize that different product behaviors drive different manufacturing architectures, a fact that affects equipment layouts, scheduling, planning and control methodologies, the number of parts to Kanban, Kanban quantities, the focus for continuous improvement, etc. Matching the product-demand behavior with the appropriate manufacturing architecture allows for the most effective performance of products to the customer.

- Select cell leaders who have responsibility, accountability, and authority (RAA) for the implementation and ownership for performance of the cells after deployment.
- Establish an effective TPM program for a focused factory/cell, critical when manufacturing equipment is involved.
- Keep designs for one-piece or small-batch flow cells simple for visual and conceptual understanding.
- Involve suppliers and manufacturing equipment suppliers in the design of Kanban and TPM programs to increase commitment to the process changes.
- Keep all shifts in a multi-shift environment involved, or gains will be suboptimal.

Testimonials

"Lean manufacturing is a powerful tool, not only for achieving performance gains unheard of in traditional systems, but also for giving employees tools that improve morale, the team environment, and a sense of accomplishment." —*Lean Steering Committee*

"Do not underestimate the power of communicating lean accomplishments to your customer ... it tells them you are controlling your costs and displays your commitment to organizational excellence." —*General Manager*

"It takes more than techniques to drive this kind of change; it takes leadership." —*Factory Manager*

GLOSSARY

Glossary

ABC material handling. The segregation of material based on replenishment lead-time, value, and part complexity. This is done to align planning and control approaches with certain types of parts for best utilization of resources. Not all parts are created equal.

Autonomation. Offers the ability to separate man and machine, because such equipment has the capability to automatically shut down when it detects a defect or abnormality. The machine stays shut down until a human being intervenes, solves the problem, and starts the machine again.

Back flushing. The deduction from inventory records of parts consumed in an assembly when the item is either booked into finished goods or sold.

Block layout. A high-level view of the factory, where square footage has been allocated, or "blocked," for specific areas. A general description of what will happen in the area is understood in order to assist with the development of material and information flow in the future design.

Cell (product cell). A clearly focused entity with the assigned resources necessary for it to control its own operational performance and satisfy customer requirements for its given products.

Cell layout. A graphical representation of the equipment/processes in a cell, typically in a U-shape, with both the operator and material flow displayed.

Cell leader. The individual selected to lead the day-to-day activities within a cell. It can be either a direct or indirect labor employee, depending on the level of cell complexity, types of decisions to be made, and capability of the workforce.

Cell team work plan. A documented schedule (calendar) of activities for the week within a cell. It lays out the game plan and provides a common understanding for all cell members as to what events should take place each week.

Cell workload analysis. An assessment of the effect of workload on equipment and processes in the cell to assure capacity to build future requirements; includes an analysis of the product demand behavior.

Communication plan. A structured process by which communication is to take place throughout the organization. It includes a definition and description as to what message will go to whom when and by what method.

Complex mix production scheduling. The same as Heijenka. The establishment of a level demand pattern sequence based on the mix of repetitive orders from the customer. For example, if demand was for 100 A units, 50 B units, and 50 C units, then the Heijenka pattern would be A, B, A, C, A, B, C, B, … .

Concept design. The first stage of the future state design phase. Concept design establishes the high-level view of what the operation will look like when the lean program is implemented. It provides the foundation for detail design.

Continuous improvement tools. Very simple tools that can be utilized by all employees to identify and eliminate waste in their process (five whys, histograms, cause-and-effect diagrams, frequency charts, Pareto diagrams, etc.).

CpK. An index measure of the capability of a process to consistently produce parts. It compares the process width (standard deviation) with the specification width and location.

Cross-training. Employees in a process being trained to perform multiple steps within the process, preferably all the steps.

Current state gap phase. The second phase in the lean manufacturing program, it is designed to capture current operational performance, to lead to an understanding of the major operational processes as they are today, and to identify root causes as to why performance is what it is.

Customer/supplier alignment. Documenting and understanding all the customer and supplier relationships that exist for part flow in the factory. It involves identifying each part and recording where it comes from and who it goes to in order to establish clear customer/supplier alignment.

Cycle time (operational). The time required to complete one full cycle of an operation. An operation is a subset of a process.

Cycle time (process). The time required to complete one full cycle of a process, made up of several operations.

Design daily production rate. The production rate developed in order to satisfy customer demand. It takes into account the customer forecast and variations in that forecast. The cell is designed to produce at that rate for a given time frame.

Detail design. The second stage of the future state design phase. Detail design analyzes what each individual cell requires for implementation during the Kaizen events — items such as takt time, equipment, demand mix, potential layout and staffing, routing analysis, etc.

DFMA (design for manufacturing and assembly). A product development approach that involves multiple functions concurrently throughout the development process to ensure all requirements are captured. It also focuses, through the use of good lean design practices, on designing a product that is production friendly with a view toward reducing recurring total costs.

Exit criteria. Quantitative and qualitative measures that are visible and can clearly show that success has been achieved. Examples of quantitative goals would include 98% on-time delivery, manufacturing lead-time of 2 days, productivity of 89%. Examples of qualitative goals would include having all A parts on Kanban, documentation of operating rules, 5S checklists, communication boards, training matrix, posted metrics, etc.

Finished-goods variation. A calculated level of finished goods based on demand variation and service level required. This finished-goods inventory is usually used for products utilizing Kanban replenishment with zero customer tolerance on delivery.

Five Primary Elements. A design and implementation approach that represents five primary facets of lean manufacturing. An approach that asserts that all facets are required in order to support and sustain a solid lean manufacturing program.

5S (housekeeping). A structured, five-step approach to housekeeping that engages both management and employees in the process. It is a matter of sifting, sorting, sweeping, standardizing, and sustaining the work environment.

Flex-fence demand management. A planning and control technique whereby customer demand is released to the cells through a set of operating rules agreed upon by marketing and manufacturing.

FMEA (failure modes and effect analysis). A technique whereby risks in the process are analyzed for potential failure based on their effect and the required function of an item.

Future state design phase. The third phase in the lean manufacturing program, it is split into two stages. The first is concept design, and the second is detail design. In addition, this phase includes the implementation plan, transition strategy, and plant communication for the program rollout.

Graphic work instructions. A graphical representation of work instructions including work sequence, work content, verification checks, and source inspections.

Holistic manufacturing. A view that there is interconnectivity and dependency among the Five Primary Elements and that each element is critical and required for the successful deployment of a lean manufacturing program.

Hoshin planning. A strategic decision-making tool that focuses company resources on a few (three to five) critical initiatives within the business and aligns these initiatives from top to bottom throughout the organization via specific goals, project plans, and progress reporting.

Implementation plan. The schedule of events for implementing the lean manufacturing program. It includes a sequence of Kaizen events, deliverables, RAA, duration, etc.

Kaizen event. A time-boxed set of activities carried out by the cell team during the week of a cell implementation. These activities include training, planning, design solutions, deployment, documentation, demonstrating performance, etc. The Kaizen event is the implementation arm of a lean manufacturing program.

Kanban. A demand signal from the customer, the authorization to begin work. It controls the level of work in process and lead-time for products. It facilitates immediate feedback on abnormalities.

Lean assessment phase. The first phase in the lean manufacturing program, it covers the initial assessment of the level of leanness of the business. It gathers external information to establish design criteria and determine market opportunities.

Lean manufacturing audit. The result of reviewing a cell implementation to provide feedback through a standard scoring process to indicate the level of deployment achieved.

Lean road map. The clarified statement, understood by all those involved, of the overall direction and steps or phases required for a particular lean manufacturing program.

Level loading. Designing a level load of demand for a given cell in order to accommodate the mix of products required for that cell (based on product volume and work content).

Line stop. Authority given to an operator to shut down the line and not produce any more product if a defect is found in the process.

Loading chart. A chart used in conjunction with takt time to establish workload balance for the work content elements of a given cell and its product mix.

Logistics element. The element that provides a definition for operating rules and the mechanisms for planning/controlling the flow of material.

Lot size splitting. Dividing a lot into sub-lots to accommodate simultaneous processing of an order.

Make-to-order production. A production architecture where products are made after the receipt of a customer sales order.

Manufacturing flow element. The element that addresses physical changes and design standards to be deployed as part of the cell.

Manufacturing lead-time. The elapsed time between when an order is released for production and the item is delivered into finished goods.

Manufacturing strategy. A collective knowledge of the business that contains current competitive advantages and weaknesses, identifies market opportunities, and includes the associated manufacturing objectives necessary to align with these opportunities.

Material planning/control. The operating rules and systems support used for planning and controlling the flow of material to, through, and from one cell to the another.

Material pull (inter-cell). A pull system for replenishing material within a cell.

Material pull (intra-cell). A pull system for replenishing material between cells.

Metrics element. The element that addresses visible results-based performance measures with targeted improvements and team rewards and recognition.

Milestone plan. A tool that identifies major segments of a project, the time frame, sequence of major events, and associated management debriefs.

Mix-model manufacturing. The ability to produce any product, any quantity, any time in order to respond to customer demand on a daily basis; designing a manufacturing cell that can produce any mix or volume of products on any given day.

Muda. Japanese word for waste, or non-value-added.

Non-repetitive Kanban. A Kanban that is used for one-off or low-volume products. It is introduced into the manufacturing process when there is a specific demand for a product. The signal is sent to the supplier for a quantity to fill the demand. After it has been consumed, it is taken out of the replenishment cycle until it is needed again.

OEE (overall equipment effectiveness). A function of scheduled availability × equipment productivity × process yield; used to understand the effectiveness of equipment.

Off-loading. Sending work to an outside supplier for a specific operation or set of operations due to a short-term capacity deficit.

One-level BOM. All component parts are at the same level in the bill of material, with no sub-assemblies, no "goes into" relationships, no lead-time offset, no structured BOM.

One-piece flow. Producing one part at a time at an operation and passing it on to the next operation after having received a demand signal.

Operating rules. New documented rules for operating the cell as designed (Kanban card system, capacity loading to 90%, incoming/outgoing material handling, workable work, recording setup times, daily equipment checks, line stop, etc.).

Operational roles and responsibilities. Documented expectations for individual positions describing what they are accountable to accomplish, specific duties to be performed, to whom they report, boundary of responsibility, direct reports, etc.

Organization element. The element that focuses on the identification of people's roles and functions, training in the new ways of working, and communication.

Pareto. The concept that a small percentage of a group has the most impact.

Poka-yoke. A mistake-proofing device or procedure used to prevent defects from entering a work process.

Policy deployment. See Hoshin planning.

Process control element. The element that is focused on the monitoring, controlling, stabilizing, and pursuit of ways to improve the process.

Process matrix. The graphical representation on a grid, with the manufacturing process across the top and part numbers down the side. Part flow is drawn inside the grid and used to reveal patterns of commonality, resource consumption, and reverse part flow.

Product-demand behavior analysis. The segregation of products into one of three categories (runner, repeater, and stranger) based on their product-demand behaviors.

Product-focused multidisciplined team. A team of people representing various functions within the organization, all of whom are focused on improving the end-product performance of a given set of products, no matter how many departmental lines those products cross.

Product grouping. The segregating of end-product demand items (SKUs) in groupings, based on defined criteria.

Product/quantity assessment. The P/Q analysis tool looks for natural breaks for product groupings by sorting the gathered data and determining a fit for product cells by their associated volumes and the product alignment characteristics.

Project charter. A tool that defines and clarifies management's expectations in regard to the purpose, objectives, and expected outcome of a project. This document must be agreed to and signed off on by all parties before a project can begin.

RAA (responsibility, accountability, authority). Implies complete ownership for a deliverable, or a process, or a performance outcome. An individual (one person) is answerable for all aspects of this assignment. This person may delegate tasks but does not share the rose that has been pinned to his or her lapel.

Rate-based schedule. Used to establish the production quantity for rate-based products in a given cell. It is determined by establishing a daily build quantity from both forecasted and booked orders, which then becomes the work schedule for the cell.

Repeater. These products have significant variety and will usually be produced across resources that are not dedicated to a specific flow line. Due to the lower volume amounts, variable order frequency and/or high variability in operational routings, these product-demand patterns will have to be managed as mix-model product and will require more production control support than a runner type of product.

Routing analysis. The categorization of products based on their process flow, work content, and volume to determine the most effective way to manage them in a cellular manufacturing environment.

Runner. These products are ordered in high volumes frequently from customers and have relatively stable demand patterns. The are often managed as rate-based products and dedicated to specific cells.

Segregated production scheduling. The grouping of products around constraints (e.g., changeover); for example, all A products are scheduled to run on first shift, while B and C products are run in sequence during the second shift due to a 2-hour changeover time between mixes.

Service cell. In contrast to a product cell, a service operation is focused on turnaround time and delivery reliability to the customer. Service cells do not have RAA for products but are held accountable for their performance to product cells.

SIPOC (supplier-input-process-output-customer). A process-mapping methodology used to capture a process, its outputs, and the associated inputs that triggered the process, in addition to identifying the customer of the output and the supplier of the input. It also collects information about the process, such as lead-time, volume, delivery, quality performance, etc.

SMED (single-minute exchange of dies). A structured improvement methodology for reducing changeover downtime on equipment to less than 10 minutes.

SPC (statistical process control). The use of statistics and data gathering to monitor process output and to control the quality of the process.

Standard work. Documentation of the agreed-upon, one best way to produce a product. It serves as the communication, training, and process improvement tool for the cell. It can include such information as cycle time, takt time, designed level of work in process, operator flow sequence, material flow sequence, staffing, etc.

Stranger. These products are the miscellaneous items that are being produced within the plant as one-off items or have a very low-volume or infrequent (once per year) demand pattern. These items are usually best managed through MRP and can be segregated from the rest of the factory.

Takt time. The rhythm or beat of demand for the cell. It represents the rate of consumption by the marketplace and is based on the scheduled time available for the cell divided by the designed daily production rate for the cell.

TPM (total productive maintenance). A structured approach to equipment maintenance involving operators, maintenance personnel, and management, all of whom have specific roles and responsibilities to eliminate unplanned downtime on equipment.

Transition strategy. Identification of specific actions required to support the implementation of lean manufacturing through Kaizen events with minimal impact on existing production (build ahead, bleed off inventory, prep work, etc.).

Transportation pipeline Kanban. Used for A-type parts that are expensive and complex, with long lead-times. The method involves filling the pipeline with constantly flowing Kanbans, each with a certain number of days' demand that results in a specific number of Kanbans in the system. The Kanbans are held and released from designated points in the supply chain so as to minimize the replenishment time to the next customer.

Visual control. The aspects of lean manufacturing that support line-of-sight management (e.g., cell name signs, painted floors, marked POU areas, performance metrics).

Volume matrix. A grid that has the manufacturing process across the top and part numbers down the side. Part-number volume, in units and hours, is applied to the work content times (from the work content matrix) to segregate high- and low-volume products and determine the degree of variation and impact on the cell design.

Work content matrix. A grid that has the manufacturing process across the top and part numbers down the side. Part-number work content for manhours, machine time, and setup time are loaded to understand variation from part to part and process to process.

Workable work. A process to verify the availability of work elements identified as being necessary for a job to go into production.

Workload balancing. Shifting the work content elements between operations in order to balance the workload for the cell to takt time.

REFERENCES

References

1. Belcher, John G., Jr., *Productivity Plus: How Today's Best Run Companies Are Gaining the Competitive Edge*, Houston, TX: Gulf Publishing, 1987.
2. Cokins, Gary, *Activity-Based Cost Management: Making It Work — A Manager's Guide to Implementing and Sustaining an Effective ABC System*, Chicago, IL: Irwin, 1996.
3. Conner, Daryl R., *Managing at the Speed of Change: How Resilient Managers Succeed and Prosper Where Others Fail*, New York: Villard Books, 1992.
4. Copacino, William C., *Supply Chain Management: The Basics and Beyond*, Boca Raton, FL: St. Lucie Press, 1997.
5. Costanza, John R., *The Quantum Leap: In Speed to Market*, Englewood, NJ: J-I-T Institute of Technology, 1995.
6. Gunn, Thomas G., *Manufacturing for Competitive Advantage: Becoming a World Class Manufacturer*, Cambridge, MA: Ballinger Publishing, 1987.
7. Hay, Edward J., *The Just-in-Time Breakthrough: Implementing the New Manufacturing Basics*, New York: Wiley, 1988.
8. Hayes, Robert H., Wheelwright, Steven C., and Clark, Kim B., *Dynamic Manufacturing: Creating the Learning Organization*, New York: Free Press, 1988.
9. Henderson, Bruce A., and Larco, Jorge L., *Lean Transformation: How To Change Your Business into a Lean Enterprise*, Richmond, VA: The Oaklea Press, 1999.
10. Hill, Terry, *The Essence of Operations Management*, New York: Prentice-Hall, 1993.
11. Hunt, V. Daniel, *Process Mapping: How To Reengineer Your Business*, New York: Wiley, 1996.
12. Imai, Masaaki, *Gemba Kaizen: A Commonsense, Low-Cost Approach to Management*, New York: McGraw-Hill, 1997.
13. Ingersoll Engineers, *Making Manufacturing Cells Work*, Dearborn, MI: Society of Manufacturing Engineers, 1992.
14. Mahoney, R. Michael, *High-Mix Low-Volume Manufacturing*, Englewood, NJ: Prentice-Hall, 1997.
15. Schonberger, Richard J., *Japanese Manufacturing Techniques: Nine Hidden Lessons in Simplicity.* New York: Free Press, 1982.

16. Schonberger, Richard J., *World Class Manufacturing: The Lessons of Simplicity Applied.* New York: Free Press, 1986.
17. Schonberger, Richard J., *World Class Manufacturing: The Next Decade.* New York: Free Press, 1996.
18. Shingo, Shigeo, *A Revolution in Manufacturing: The SMED System.* Portland: Productivity Press, 1985.
19. Shingo, Shigeo, *A Study of the Toyota Production System from an Industrial Engineering Viewpoint*, Portland: Productivity Press, 1989.
20. Shingo, Shigeo, *Non-Stock Production: The Shingo System for Continuous Improvement*, Portland: Productivity Press, 1988.
21. Shingo, Shigeo, *Zero Quality Control: Source Inspection and the Poka-Yoke System*, Portland: Productivity Press, 1986.
22. Shingo, Shigeo, *The Sayings of Shigeo Shingo: Key Strategies for Plant Improvement*, Portland: Productivity Press, 1987.
23. Shirose, Kunio, *TPM For Workshop Leaders*, Portland: Productivity Press, 1992.
24. Tobin, Daniel R., *Re-Educating the Corporation: Foundations for the Learning Organization*, Essex Junction: Oliver Wight Publications, 1993.
25. Womack, James P., and Jones, Daniel T., *Lean Thinking: Banish Waste and Create Wealth in Your Corporation*, New York: Simon & Schuster, 1996.
26. Womack, James P., Jones, Daniel T., and Roos, Daniel, *The Machine That Changed the World*, New York: Harper Collins, 1990.

INDEX

Index

H

I

J

K

L

M

O

P

U

V

W